Imagenation

Popular Images of Genetics

José Van Dijck
Associate Professor of Literature
University of Maastricht
The Netherlands

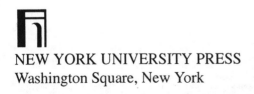

NEW YORK UNIVERSITY PRESS
Washington Square, New York

£18-50

First published in the U.S.A. in 1998 by
NEW YORK UNIVERSITY PRESS
Washington Square
New York, N.Y. 10003

This book is printed on paper suitable for recycling and
made from fully managed and sustained forest sources.

Library of Congress Cataloging-in-Publication Data
Dijck, José Van.
Imagenation : popular images of genetics / José Van Dijck.
p.   cm.
Includes bibliographical references and index.
ISBN 0–8147–8796–7 (hbk.). — ISBN 0–8147–8797–5 (pbk.)
1. Genetics in mass media.   I. Title.
P96.G45D55 1998
576.5—dc21   97–34591
                                                              CIP

Printed in Great Britain

Contents

Acknowledgements

I welcome this opportunity to express my gratitude to all institutions and people whose help has been indispensable in completing this project. I am most thankful to the Fulbright Foundation, who awarded me a fellowship to finish my research for this book in the USA. The Department of Arts and Culture at the University of Maastricht, and especially my colleagues in Literature, allowed me precious research time. Rein de Wilde and Wiel Kusters patiently read through several immature drafts of this manuscript; their comments have been extremely constructive. I would also like to thank the members of the BOTS-research group for their criticism and Wiebe Bijker for his encouragement. Lilian Essers gave me much appreciated administrative support.

I was most fortunate to spend several months as a Resident Scholar at the Center for Cultural Studies at the University of California, Santa Cruz. I would like to thank Gail Herschatter and Chris Connery for providing a stimulating intellectual environment, and Phil Stevens and Katy Elliott for offering valuable logistical support and an office that felt like home. I appreciate the time that scientists and employees of biotechnology companies spent talking to me or sending me research material, most notably Robert Sinsheimer and the public relations department at Genentech. Life in Santa Cruz would have been a lot less perfect without the warm friendship of Mary Ellen Boyle, Katherine Beiers and Woutje Swets. Craig Reinarman and Monica Casper fed me great meals, sociological theory and endless movies.

My friend and colleague Kathy Davis deserves a trophy for spending her precious Christmas break reading my manuscript; this book would not have been the same without her meticulous criticism. I am much indebted to Rick Powers, who carefully edited the final version of this manuscript, and who deepened and fine-tuned my knowledge of genetics. Ton Brouwers endured so many drafts of this book, it is amazing he still had enough patience left with me in the end; I am extremely grateful for his steady support. John Smith, my editor at Macmillan, and Ruth Thackeray have done a superb job with the final editing of this book.

JOSÉ VAN DIJCK
Maastricht

'Imagine having the tools to keep up with your imagination'

Advertisement for Biosearch Labs, 1994

Introduction

'Monsters manufactured!' exclaims Edward Prentick, the unfortunate shipwrecked man in H.G. Wells's *The Island of Dr Moreau* (1894), as soon as he finds out what the doctor has done to the imported animal population on the island where he has been stranded.[1] Moreau is a 'vivisector' who tries to modify animals into human beings, using techniques like skin grafting, transplantation of body parts and other methods of inoculation with living and dead tissue. His intervention in the course of evolution is motivated by a firm belief in the perfection of the human race through technological innovation. Doctor Moreau assures his baffled, involuntary guest on the island that he regards himself an emulator of God, since the creator has not been very efficient in his enforcement of evolutionary laws. Depending on the grade of his technology's perfection, the vivisector thinks he can upgrade and control the island's animal population. The moral of this story follows from the inevitable punishment of the arrogant scientist, as he gets killed in the uprising of animals.

One hundred years after the novel's publication, the movie version of *The Island of Dr Moreau* turns out to be a peculiar re-creation of H.G. Wells's science fiction fantasy.[2] The laboratory on Moreau's island is now equipped with microscopes, embryo bottles and DNA sequencing machines, rather than operating rooms for skin-grafting procedures and body part transplants. Instead of holding on to the 'mechanical' reassemblage of body parts, the movie-makers have updated Wells's technological apparatus to include the latest developments in genomics. The opening scene of the movie features a collage of cells, DNA helixes and enlarged microscopic pictures, symbolizing the control of humans over evolution through intervention at the molecular level. Douglas, the cinematic equivalent of Edward Prentick, realizes that his DNA has been used in Doctor Moreau's experiment to keep the hybrid creatures from regressing into an animal stage of evolution. In addition, Moreau uses the newest equipment, such as electronic implantations and hormone replacements, to control and regulate his subservient populace of engineered animal-humans.

Although the technology in this movie has been brought up to date, evidenced by the abundant visuals of DNA, the moralist tenet and ideological contention of Wells's story have been left surprisingly intact. In the movie, we are confronted with the fixed hierarchy of the earth's species in a world where God reigns over men, men over animals, and animals over lower forms of civilization. Colonialist overtones amplify the suggestion of white superiority over coloured peoples who are lower on the evolutionary ladder. While gene technology and media technology – like screen-editing and special effects – facilitate both the 'fixing' of cellular organisms and the representation thereof, technological innovations are not exploited to 'update' the underlying ideology of racism. New technological tools, evidently, do not lead automatically to a retooling of the imagination.

In the 1990s, genetics seems more 'popular' than ever. DNA-technology not only sustains large areas of biomedicine and business, but also prevails in a number of social and legal practices, and obviously takes root in cultural products. We can read in a science journal how, in the near future, blue jeans may be manufactured with the help of blue genes; companies are busy developing genetically engineered cotton plants to replace the toxical blue dyes used in most blue jeans today. A business magazine features an article about the 'gene blues': small biotech companies are going bust because there are too many 'copycats' on the market, and investors are getting impatient. In common parlance, expressions like 'genetic obligation' have replaced the former 'family bonds', and on the front page of the newspaper, we can read that 'politicians and movie stars spring from the same DNA', referring to their shared talents to put up performances.[3] And movies like *The Island of Dr Moreau* and *Jurassic Park* imagine how we, with the help of DNA-technology, can manipulate the evolution of species.[4]

The permeation of genetic thinking seems ubiquitous, affecting virtually all aspects of late twentieth-century life – manufacturing, business, law, politics and the entertainment industry. There is no denial that the advancement of genetic technology has left its mark on everyday practices, and has led to what many have called the 'genetic turn'. But despite the apparent popularity of genetic thinking, the advancement of the new genetics has not resulted in total acceptance or univocal acclamation. Genetics has always been a controversial field, and the advent of ever newer technological tools has not effaced that public concern. Since time immemorial, the potential of human intervention in the course of evolution has elicited profound fears as well as hopes. Every new technology in the field of genetics changes the outlook on humans' control over health, disease and reproduction. And every new tool has triggered fantasies of how these technologies can be used and abused as instruments of control. The dissemination of genetic knowledge is not uniquely contingent on the advancement of science and technology, but is equally dependent on the development of images and imaginations. 'Imaginary tools' are crucial assets in the dissemination of genetic knowledge, as they are used to shape this science's public face.

In this book, I will examine the role of images and imagination in popular representations of the new genetics since the late 1950s. Popular representations have proliferated through a number of different discourses. For the purpose of this study, I have concentrated on non-fiction books written by scientists and/or journalists, general interest magazines, popular science magazines, (science) fiction, advertisements and public relations material. To limit the broad range of related fields and practices, I will focus solely on human genetics, excluding genetic modification of plants or animals. Hence, *The Island of Dr Moreau* and *Jurassic Park* will not return in this book, however interesting their re-imaginations of contemporary gene technologies. Applied to humans, genetic engineering and gene therapy have generated an array of inquiries into their feasibility and permissibility. Although many of these inquiries take on the form of fiction or speculation, they have equally informed scientific non-fiction and media reports.

Popular representations of genetics do not simply reflect the gradual transformation of genetics from a suspect branch of research into a thriving medical field, but show that genetics has been a continuous site of contestation. In Chapter 1, I will explain that popularization is a multi-layered process: a contest to define the hegemonic meaning of genetics, as well as a contest over its representation. Special interest groups and professional groups mobilize images and imaginations in their debates over the meaning of genetics. In order to differentiate various aspects in this process of image-making, I propose to look at genetics as a 'theatre of representation'. This metaphor will serve as an analytical tool to dissect the display of competing images at various moments in the representational history of the new genetics.

Chapters 2 to 6 will take the reader through four 'stages of imagenation'. Chapter 2 starts with the introduction of the 'new biology' paradigms in the late 1950s and 1960s, when genetics was primarily debated in terms of morality. Molecular biologists had a hard time establishing a positive image for their 'new' field in a cultural climate that was inundated with lingering shadows of eugenics' unfavourable past. In Chapter 3, we will see how, in the 1970s, a debate erupted over the environmental safety of genetics. Stories of escaping DNA-strings dominate the media, as activists and scientists dispute each others' assessments of recombinant DNA's potential harm to the environment. A rapidly emerging biotechnology industry in the 1980s sets the stage for the contestation described in Chapter 4. Mechanical and business images of genes precipitate a boom in bioengineering – the era of 'biomania' – and these images played a key role in raising private capital. Chapter 5 will centre on the popular images generated by the Human Genome Project. The new concept of genome 'mapping' gives rise to a broad range of images and imaginations, some of which reflect the revolutionary changes in genome technology, while others re-invoke conventional interpretations of the genetic code. Even in the era of 'biophoria', human genome discourse is met by 'biocriticism', and Chapter 6 shows how detractors of the Human Genome Project take on the inadequacy or inaccuracy of pervasive genome images.

The aim of this book is not to provide a historiography of popular images of genetics, but rather to explain the recurring paradox between changing technologies and persistent representations thereof. As elucidated in Chapter 7, such incommensurabilities can only be accounted for if scientific practices are examined in conjunction with representational practices and in the context of cultural practices. Ideological tenets have always shaped the cultural forms and (narrative) conventions by means of which we make sense of new developments in science and technology. Popular representations of science are thus both reflections and constructions of technologies, always part of a struggle to make new ideas accepted. It is this process of image construction that will be at the heart of this analytical inquiry.

1 Popular Images of Genetics

THE TRANSFORMATION OF GENETICS

The history of genetics is commonly viewed as a succession of scientific milestones, starting with Darwin's theory of evolution, via the formulation of Mendelian genetics and the 'discovery' of the double helical model, ending with the latest breakthrough in gene therapy or the public disclosure of another gene-of-the-week. A fascination with human genetics has been allegedly on the rise since the mid-nineteenth century, or 'ever since Darwin', and has become a central focus in medicine and biological research.[1] The vindication of what is called the 'new' genetics is assumed to start with the discovery of the double helix in 1953, and finds its provisional apogee in the development of the largest government-funded project ever in the field of biology: the Human Genome Project – a three billion dollar effort to map the entire genetic make-up of homo sapiens before the year 2005. Molecular biology has managed to shake off its compromised association with former eugenics, and has become a generously funded and well-respected area of research, in which nothing less than a 'revolution of medicine' is at stake. Genetic thinking has apparently dethroned social determinism as a dominant social force, and has percolated into all aspects of society and culture. A growing interest in DNA as an explanation for human disease and deviant behaviour has culminated in what sceptics have called 'the genetic turn' or 'the geneticization of science and society'.[2]

In state-of-the-art reports on the advancement of genetic research, we are usually confronted with the teleological order of a scientific theory that led to new practices, which in turn resulted in their general acceptance by a large group of people. A gradual ascendance in the public esteem, in this line of reasoning, is a corollary to genetics' technological advancement. On the face of it, the story of geneticization appears one of gradually evolving successes and triumphs, ending in widespread application of scientific paradigms. As an ordinary tale of scientific progress, the story of geneticization could be told as the simple transformation from small into 'big science'. Yet such simple plotting would be far too reductive as a description of an intricate formative process. Despite its current 'popularity', genetics has never been – and probably never will be – unchallenged. Ever since the eugenics movement's heyday in the 1920s and 1930s, public perception of genetics has been clouded by normative considerations, and has been disputed by various groups.[3] Rather than a straightforward advancement towards public embrace, we find a continuous contestation of its meanings.

Depending on whom you ask, attempts to clarify the meaning of 'genetics' will result in a palette of diverging definitions. Most scientists will point at the 'scientific' meaning of genetics as a branch of biology that deals with the principles and mechanisms of heredity, and with the genetic properties or molecular constitution of an organism. Others might describe it as a scientific field that works on the development of scientific methods and technologies that permit direct manipulation of molecular material in order to alter the hereditary traits of a cell. Yet another expert might define genetics as the scientific paradigm on which the development of a revolutionary (bio)technology is based. Ask any layperson, however, and you can expect a wide variety of descriptions: genetics is called the 'science of life'; a potential cure to many hereditary diseases; a scary practice that involves tinkering with the essence of human life; the professional home town of genetic engineering; or the promise for genetic therapy. Carved into these descriptions and definitions are evaluative remarks – signs of the highly controversial nature of the defined object.

Throughout the past fifty years, genetics has raised questions that are of widespread public interest and concern. Since the gene has come to be considered the basis of life, replacing the cell or the organism as the locus of vital activity, there has been a continuing public discussion surrounding the potentialities and permissibility of various genetic technologies. Every new development or discovery in the field of molecular biology has triggered an array of inquiries into the legal, moral, ethical, psychological and social effects of genetic engineering. Genetics, perhaps more than any other scientific discipline, has proliferated as a *public issue*, of a controversial nature.[4] Public discussions on the 'new genetics' have been infused with a mixture of hope and fear. On the one hand, genetics promises a cure to congenital disease, while on the other it raises the spectre of scientists tinkering with human bodies at the molecular level. Both configurations sprout from a view on DNA-molecules as the 'essence of life'. Molecular genetics was formerly known as 'genetic engineering' – a cluster of macro-manipulations of the reproductive or hereditary level, some of which, like cloning, have little to do with genetics, but came to stand out as the apex of public fear.[5] The potential applications of genetic modification or genetic therapy range from curing congenital defects by inserting 'healthy' genes in the body, to removing the defect from egg and sperm cells and engineering wholesale changes in the human gene pool. The new biology has generated discourses of fear and hope, of manipulation and modification, and of scientists being out of control or in control.[6] What makes genetics concurrently so frightening and seductive?

The kinds of fear generated by the new genetics are various in nature. The prospect of genetically engineered DNA-strings irreversibly affecting the course of evolution, causing 'biological pollution' and disturbing evolutionary balance, has repeatedly come up in discussions. Another major concern relates to power: Should humans take control over the evolution of their own race? Genetic engineering in plants and animals is often pointed at as a precursor for future development of designer babies and customized genetic material. The power of a small group of in-crowd scientists who have the knowledge to define what the future of the human race will look like also raises anxiety. Potential abuse of quantified genetic information elicits worries about discrimination on the basis of genetic predisposition. And, on a more philosophical level, DNA-manipulation and genome research stir up profound agitation over the integrity of the human body and the corrosion of human identity. The genetic make-up represents the deep structure of inner life that has traditionally been met with awe, mystification and veneration.

Part of genetics' appeal is its promise to cure congenital disease, or even eliminate disease from the roots of human reproduction. Since the 1950s, we can notice a clear shift in the public perception of genetics, from an obscure scientific paradigm (genetic engineering) into a preferred solution to a pressing medical problem (genetic therapy). The cause of ever more diseases is purportedly fixed in the genes. This attribution to genetic causes is not restricted to so-called single-gene disorders, such as cystic fibrosis and Tay Sachs disease, but seems to be extended to a number of multi-factor diseases, and even to behavioural deviancies. Naturally, the idea of genetics as a remedy for congenital diseases raises different concerns from the prospect of genetics as a remedy for behavioural deviance.[7] Yet through media and other public channels, we are confronted, in one and the same breath, with genes for single-gene diseases, breast cancer, alcoholism, violence or manic depression. Public affairs magazines have shown a remarkable taste for this gene-discovery hype, as they seriously report on the gene for infidelity, the gay gene and the gene for criminality.[8] Besides the equation of disease and deviant behaviour, these gene-discovery claims imply the promise that the localization of a gene will automatically lead to a cure. No one can reasonably dispute the advantages of a new technology that cures people who suffer from congenital diseases. Yet the emphasis on curative techniques like gene replacement therapy augments the emphasis on genetic components as the cause of all disease and deviance, while concurrently enhancing the need for genetic therapy.

Even though only a tiny percentage of the population is endowed with diseases that can be unambiguously ascribed to genetic causes, these causes have increasingly moved to the centre of public attention. But explaining a mounting interest in genetic causes as the self-evident consequence of major advancements in molecular biology, would be as unwarranted as explaining the surging interest in genetic therapies and genetic testing as the result of a sudden compassion for victims of rare inherited afflictions. Public interest in a scientific issue never arises 'naturally' from scientific developments, or from an increase in victims affected by a disease. In order to gain public recognition, a scientific issue needs a sense of urgency – a potential for human *drama*. Science that entails a solution to a pressing human need has a better chance to appear on the political priority list than science that does not.

Medical issues, *par excellence*, have a highly dramatic presentational impact. Health and disease have long topped the list of people's quotidian concerns, and medical science has become paradigmatic for the popular representation of science.[9] Not every medical issue is equally dramatic, however. Diseases like cancer have waxed and waned in the public's attention. The problem of infertility – as well as the proposed medical remedy for this 'serious epidemic' – became a highly publicized issue in the 1980s.[10] Public recognition of medical issues as urgent medical problems is not determined by the number of people afflicted by a specific type of disease, but often by the possible effect that disease may have on the moral or social order. AIDS, for instance, was initially cast as a disease of homosexuals, as 'their' sexual mores assailed the puritan heterosexual morale.[11] Although many more people in the world are suffering from malaria than from malfunctioning fallopian tubes, the problem of infertility exhibits a greater sense of urgency and apparently has more dramatic impact. Obviously, the way in which genetics is framed as a medical drama adds in a major way to its 'popularity'.

Why do some diseases and favoured remedies move to the spotlight of public attention at one particular moment, while other diseases that are much more common remain obscure? Why and how do some scientific theories gain popularity? Of course we can never examine the cause, nor assess the exact magnitude, of a change in popularity or public perception. Yet the question why certain scientific paradigms or medical problems come into vogue at one time or another, brings up important questions about the nature of popularity and the impact of popular representation on public perception.

THE POPULARIZATION OF SCIENCE

'Popular', in the context of science, is a slippery term. First, it means science 'relating to the general public' or 'suitable to the majority'. Second, the term pertains to science that is 'easy to understand'. And last, the term refers to scientific issues frequently encountered or commonly accepted. The term 'popularization' wavers accordingly between 'rendering prevalent among a general audience' and 'presenting science in a generally understandable form'. But the two meanings are commonly regarded as two sides of the same coin: by presenting science in a generally understandable form, its recognition and acceptance is assumed to increase proportionally. Public understanding of science, some have argued, is too easily equated to the public appreciation of it.[12] To popularize science commonly means to 'sell' science to a general audience – to make as many people as possible 'buy into' a particular scientific theory or practice.

The dominant view on science popularization is based on similar assumptions as the straight arrow theory of scientific progress: public acceptance of science is simply a matter of overcoming resistance. Resistance to scientific paradigms or ideas might come from special interest groups, who fuel the discourse of concern, but can also be expected from professional mediators such as journalists. Just as special interest groups supposedly form an obstacle to the dissemination of science, journalists and media are thought to pollute or distort science.[13] And beyond opponents and journalists there are other 'hurdles' to overcome: government regulations, political restrictions, social norms and ethical taboos. In this perspective, the main objective of science popularization is its diffusion, or its acceptance by as many people as possible.[14] According to the 'diffusion model', scientific and technological innovations – usually generated by a few great men or geniuses – are gradually distributed among a large, ignorant audience of laypersons, who have to be initiated into the wonders of science, and render them more appreciative and enlightened.[15] Society is made up of groups which resist, accept or ignore both facts and technological innovations. The outcome of the diffusion model is a division between groups of scientists who promote science, and social and professional groups who resist acceptance, sustaining the idea that there is an obvious split between science, technology and society.

Various critics have convincingly argued that the 'diffusion model', however outmoded and flawed, remains a prevalent way of thinking about science popularization.[16] Popularization or public acceptance of a particular type of science, far from being a linear process of overcoming 'resistance', is a process of *translation* and *negotiation*. Genetics is a locus of contestation, in which special interest groups and professional groups are involved in defining the hegemonic meaning of genetics. Rather than being 'fixed' categories, these groups constantly shift alliances, try out new associations, displace and redefine goals of science, and invent new strategies. Since the 1950s, scientists have been met by environmentalists, feminists, sociobiologists, political activists, ethicists, business executives, religious representatives, politicians and other groups who have all contributed to genetics' public meaning. These groups engaged in changing discourse coalitions; they looked for strategic alliances and contested each others' claims to knowledge. Scientists have – sometimes reluctantly, sometimes eagerly – teamed up with other groups to strengthen a specific definition, and have altered their strategies and redefined their goals if necessary. If we look at science as a process of translation or negotiation, scientists do not form a homogeneous group, taking a unified stance on genetics; and neither do feminists share a single view on this topic. Groups are by definition heterogeneous and unequal in power, as scientists obviously have more social power than environmentalists or feminists.[17] Despite their powerful positions in the discursive hierarchy, though, scientists have never had absolute authority over the interpretation of knowledge, and thus had to look for effective strategies to propel and defend a specific interested position on genetics.

The popularization of genetics can also be construed as a site of contestation over science's representation. Popular representations of science are commonly viewed to be generated by non-scientists – journalists, fiction writers and others. Scientists are generally regarded as 'producers' of scientific knowledge, whereas journalists, novelists or fiction writers are ascribed the function of 'distributing' expert knowledge and creating popular stories. Popular media accounts of science, in line with this model, are viewed as attenuated truths at best or distortions at worst. The idea of media distortion is based on a hierarchical model, in which scientists take a higher position than journalists, but where journalists have absolute power over the popular representation of science. Each group claims a distinct authority in the shaping of public knowledge, thus demarcating their professional auhorities and responsibilities. Whereas scientists are commonly thought to command the fortress of factual knowledge, journalists are supposed to reign over the domain of factual representations. Public relations managers are purportedly in charge of the creation of favourable public images, and fiction writers control the domain of the imagination. In contrast to this hegemonic view, I will argue that we need to view the 'mediation' of science as a chain of interactions and alliances between various professional groups, who are not merely facilitators or manipulators of expert knowledge, but who are themselves active participants in a public definition of science. The mutually defined roles of journalists, scientists, public relations officials and fiction writers are never defined once and for all, but they are subject to delineation. The popularization of science can be characterized as a process of professional discursive pickpocketing, ally recruiting and boundary negotiation.[18] Popular representations of genetics are not simply constructed 'inside' journalism or popular culture, or 'outside' science, but the shifting boundaries between those domains add in a major way to the signification process.

Constructivist and related approaches to science have abundantly shown that social developments are inherent in technological innovations, and that scientists as well as other actors – alongside machines, laboratories and literature – are involved in fact-producing processes.[19] The gradual percolation of genetic theories in all aspects of society and culture is not a self-evident consequence of scientific advancement but the result of an intricate process of interaction and negotiation. Evidently, the production of 'facts' – 'statements with no modality and no trace of authorship' – is at stake in any attempt to translate genetic claims to a general audience.[20] Yet besides a 'fact-producing process' science is also an 'image-producing process'. Of course the production of images can never be completely separated from the production of facts, but with an eye on the highlighted popularization of genetics, an artificial distinction of images from facts may be justified.[21] When we talk about dissemination of knowledge, we can obviously discern logical arguments that are mobilized in defence of a particular definition or evaluation of genetics. But beyond the invocation of *logos*, there are frequent appeals to *pathos* and *ethos* to persuade a general audience of the validity of a specific interpretation.[22] Popular appeal often takes shape through the evocative use of mental pictures or compelling stories – or through *images* and *imaginations*. Popularized science thrives on the use of images, maybe even more than on logic and arguments, but the production and distribution of popular images is seldom taken as a prime object of serious critical inquiry.[23]

GENETIC IMAGES AND IMAGINATION

The term 'image' is as ambiguous as the word 'popular'. Images may infer tangible or vivid representations, as they are employed to call up mental pictures. Image also means 'popular conception', referring to fixed ideas or portraits engraved in our collective consciousness. Fixed images may signal emotions, such as fear or horror, but may also call up favourable or desirable associations. Many organizations, companies or entire scientific disciplines actively strive to inculcate a favourable 'imago', an idealized mental picture, in the public mind. In all its variants, an image is never an exact copy of something, but an inherently slanted representation that elicits a web of associations and connotations. Images, contrary to arguments, attach *unarticulated*, *implied* information to a statement or claim. In popular science representations, images may be used to concretize an abstract scientific concept, to link up complex theories with concrete products or (un)desirable results, or to create a particular imago of a scientific discipline or a group of scientists.[24] The term's ubiquity renders it useful as an analytical concept in the fields of science, journalism, public relations and literature.

Images are important resources in the popularization and public evaluation of science: they shape the public's perception of what genetics is and what it enables. Endless translations of goals and needs take place through the formulation of images. We may find the image of a discipline tinkering with the genetic make-up of human beings being countered by images of a science that remedies cell mutations caused by nuclear radiation. Unarticulated fears of DNA-experiments posing a threat to the environment may be offset by images of genetically engineered microbes eating up oil spills. Besides shaping scientific concepts and goals, images can also be employed to associate products with desirable applications. Phone companies like AT&T, if we are to believe their advertisements, are not in the business of selling phone calls, but of 'connecting people'. In a similar way, genetics is keen on creating favourable impressions of their products, people and scientific activities. Gene therapy may be portrayed as the scientific activity that cures small kids from congenital disease, giving them 'a chance to live'. But conversely, genetic therapy may also be touted as the immoral manipulation of genes and cells. Throughout the past four decades, the popularization of genetics partly materialized via a 'translation' of its favourable or unfavourable images.

The public appeal of genetics is contingent on the shifting mobilization of images, but also on the projection of imaginations onto theories, practices, instruments or applications of genetics. Imaginary projections strategically infuse a particular ideological claim. A fear for gene manipulation can be derived from stories of clone-factories or tales of obscure men tinkering with human life. The abstract idea of gene therapy becomes much more appealing through the story of a patient with a rare immune disorder treated by a team of doctors who try to cure the victim by inserting a new gene. Imaginations are often about what we want to believe, what we project or imagine that science and technology can do for us in the future. Imagination is the lubricant that ties in ideological concepts and scientific inventions. It inspires scientists as well as the public to conceive of unexisting technology as desirable or frightening. Public appeal of a scientific idea is often based on imaginary appeal; more people will *buy into* an abstract scientific construct if it elicits a familiar or desirable image. Or, as Marcel C. Folette sums up in her account of early twentieth-century public images of science: 'In the free marketplace of ideas, there is a direct relationship between the popularity of a concept and how well it matches the audience's beliefs about that concept.'[25] Except for a site of contestation, popular images also comprise a market – a place where buyers need sellers need products. The exchange of images inherently reflects the tri-directionality of the 'marketplace of ideas'.

The anticipatory effect of imagination is not created *in* fiction or *in* science, but is produced in culture at large, as it structures both factual and fictional stories of science. Stories of genetics commonly refer to techniques that have yet to be conceptualized and developed.[26] The need for a new future technology is often based on unstated premises and collective fantasies. Until 1969, when the first human being walked on the moon, the desire for this technology was nursed by science fiction stories in which humans set out to discover and conquer other planets. Only implicitly do we find allusions to the 'need for' this adventure: an overcrowded planet earth requiring the discovery of additional living space. In a similar fashion, the explicit goal to penetrate the 'secret of life', underpinning concrete theories of molecular biology, is often warranted by the implicit desire for humans to control the quality and quantity (duration) of human lives. The fantasy of cloning may be mediated by the subconscious aspiration to prolong an individual's life, but can also be articulated as a fear for prolongation of power, or the destruction of a unique personal identity. The need for germ line therapy may be motivated by a desire to save future generations the pain of single-gene diseases, but at the same time it elicits worries about bodily integrity, self-identity and the essence of the human species. Imaginations push an unspecified set of signposts through which genetics is interpreted and imagined.

A unique and significant feature of the imaginary is that it ascribes *symbolic* meaning to science and technology. In her essay on the construction of the Underground in the nineteenth century, Rosalind Williams shows how symbolic imagination plays a pertinent role in the development and implementation of technological systems in society.[27] Exploring images of subterranean worlds, she demonstrates that the idea of a perfectly working Underground infrastructure for sanitary and transportation systems gradually superseded conventional associations of the underworld with hell and apocalypse. The imagination of the Underground, in Williams's study, is tied in with class struggle: construction of a subway infrastructure was often figured as the site of hell for lower class labourers, whereas the middle classes espoused a view of underground transit as the cleaning and ordering of the city. A large technological innovation is thus intricately linked up with public imaginations that accompany this project, as Williams concludes: 'Subterranean iconography connects the historical experience of excavation and the literary interpretation of underworlds as technological environments' (81). Symbolic meanings play a significant role in the public discussion on genetic engineering. Genetics discourse is replete with symbolic allusions and associations. Religious and mythical references abound in popular accounts, as they form powerful resources in the struggle for signification. The very idea of genetics as the 'science of life' betrays an attribution of demiurgic powers to geneticists who are concerned with secrets of life and death.

The symbolic function of the imagination proliferates most explicitly in the textual genre we call science fiction. As scenarios of future scientific or technological innovations, science fiction envelops the tension between technical and literary invention; it allows imaginary space to explore the various effects of technical creations, it calls into question our sense of the world as it is and questions the need for new technologies. The advent of bioengineering has given rise to a number of fantasies and extrapolations, from cloning to gene manipulation. Through utopian or dystopian fictional filters, the balance between the potential and the admissible is weighed. Ethical and moral implications of genetic engineering may be more poignantly assessed in fictional frames. These fictions combined – products of our individual and collective cultural imagination – actively shape popular images of genetics. In other words, they are not only reflections of collective anxieties and fantasies, but also contributions to the public meaning of genetics.

Although stories of desire and fear, projection and speculation, most explicitly surface in science fiction fantasies, they are certainly not absent in non-fiction. Molecular biologists and geneticists quite often rely on imaginary narratives to elucidate scientific achievements, or implicitly to relay the goals of their research. Telling and reframing stories is an effective rhetorical tactic when it comes to impressing the desirability or need for a specific technology in the public mind. The power of narrative is that it serves as a mode of cognition without it being itself recognizable as an interpretive framework. Just as the use of a telescope mediates the visual field, narrative mediates the concept and perception of science, rendering itself invisible in the process of narration. Every story on genetics, whether a science fiction novel or a non-fiction book, displays essential ingredients of narrative. It is through the lens of story-telling that we can perceive communal rhetorical strategies and images in the various discourses on genetics.[28]

The way in which we interpret stories of genetics depends to a large extent on the conceptual frameworks that mediate our understanding of culture in general. Narrative analysis bares those conceptual frameworks that undergird hegemonic and critical interpretations of genetics. Images and imaginations, like arguments, are mobilized to affect the meaning of genetics. Popular images are sites where simultaneously the *control over science* and the *control over representation* is at stake. Launched to compete in the public arena, images and imagination betoken the ideological and political contexts in which genetics is publically disputed. An analysis of popular images should be simultaneously an analysis of representation and its representers – the way they deploy images and imaginations to transform the meaning of genetics. Since the shaping of genetic knowledge occurs in science and journalism as well as popular culture, we should look concomitantly at the production of images and imaginations in all domains. Viewing science as an image-making process implies that popular representations are produced in and distributed through a variety of fields, instead of being exclusively contained to one.

Images and imaginations are tools used to create and push forward meanings. They are not inherently innovative or consolidating, however; their transformative force depends on how they are mobilized, and how they are picked up and disseminated by others. Narrative techniques or imaginary tools make people see things or see them differently, yet they can also be employed to confirm a pre-existing or lingering emotion. The infusion of images is always strategic, as concepts of genetics can both be confirmed by, and altered with, images or imaginations. If we view the popularization of genetics as a series of image convulsions, a number of questions come to mind. Which images appear, disappear or reappear in the shaping of public knowledge? What tenets underpin the preference for certain images? Do images and stories change along with scientific innovations? What is the role of popular stories and science fiction tales in the distribution of genetic knowledge? Although genetic technology gives rise to new imaginations, why do some images recur so persistently? Who are the people involved in the process of image construction? For whom do they speak?

The questions raised here require an analysis that pertains to various aspects of the image-making process: the production and circulation of popular images among various special interest groups, the dynamics between the professional groups involved in their mediation, and the context in which images and imaginations are exchanged. In order to make this multi-layered analytical inquiry operational, I will propose to look at genetics as a 'theatre'. I have already mentioned how the need for scientific issues to appear in the public mind as urgent, pressing problems, leads to their dramatization. Popular science can be seen as a theatre, a site of performance where dramas are staged, which allow an audience to watch and interpret its content. The use of the theatre metaphor for science is not new or original. Joseph Gusfield launched the analytical metaphor in the 1970s, and more recently, N. Katherine Hayles has adopted this term to study metaphoric networks as constitutives of meaning production.[29] In Hayles's definition, the 'theatre of representation' is a public arena in which not only scientists, but also journalists, activists, politicians and others present their own and each other's views, using an array of rhetorical techniques. These discourses, deployed by several groups, are not isolated or monopolized, but they are interdependent and mutually inclusive. A dominant meaning of science is as much the outcome of political and rhetorical negotiations in the public arena – the semantic and symbolic turf contested – as it is the result of dynamics between the groups that contest its meaning; the product and the process of meaning production are thus closely interrelated.[30]

In my analysis of popular images of genetics, I adopt Hayles's concept of science as a theatre of representation, only to refine its conceptual function. The 'science as theatre' metaphor allows the distinction of three perspectives on the process of image-making: the angle of *performance*, *production* and *context*. Viewed as a performance, we can dissect the 'theatre of genetics' as a drama in which nothing less than a hegemonic definition of genetics is at stake. Various special interest groups – actors on the stage – propel images and imagination in competition, not in contemplation. From the perspective of production, a theatre involves a number of professional groups engaged in the scripting, staging and setting of a performance, such as script writers, reviewers, public relations managers, producers and others. Translated to the 'theatre' of genetics, popular images and imaginations are equally generated by scientists, public relations managers, journalists and fiction writers, and the dynamics between these groups are inherently part of their meaning production. A third possible angle is to look at how a performance or production is embedded in its larger social, historical and cultural context. Popular images of genetics are always struck against a backdrop of general historical developments, social concerns and cultural developments. Since a distinction of these three angles is pivotal to the understanding of my critical inquiry into the popular 'imagenation', I will elaborate on each of them in more detail.

GENETICS AS PUBLIC PERFORMANCE

Looking at genetics as a performance, three main ingredients of this staged act that immediately come to mind are *characters*, *plots* and *metaphors*. The terms 'image' and 'imagination' may equally pertain to a person, a scientific field and an object. If we speak about the image of scientists, we talk about their public personae – personae who are constructed or construct themselves as characters. The image of genetics – a disciplinary field or a scientific activity – can be affected through the projection of constructed or reframed plots. The gene, finally, is most commonly explicated through models and metaphors. We will find that popular representations of genetics are articulated through the *characters of geneticists*, the *plots of genetics* and the *metaphors of genes*.

Over the years, the popularity of a scientific field has become more dependent on the public personae of its practitioners.[31] More than any other science, genetics badly needed its promoters, since the image of this new branch was far from glamorous. The label 'geneticist' aroused more than a little suspicion in postwar decades. A familiar depiction of a geneticist was that of a suspect, esoteric individual involved in incomprehensible lab experiments. In contrast to astrophysicists, who were commonly perceived as innocent star-gazers, or mathematicians, who were merely looked upon as theoretical model-makers, geneticists were perceived to be engaged in potentially life-changing experiments. Because the outcome of their research was expected to have a profound impact on society, they were evaluated, more than other scientists, in terms of good or bad. Especially in the age of 'big science' – with projects such as the Space Program taking up large chunks of research funds – geneticists and molecular biologists had to distinguish themselves from colleagues in competing fields, and hence had to work on their public image.

Evidently, the image of geneticists is partly dependent upon the image of scientists in general. The very word 'scientist' evokes diverging mental images: where one sees Nobel laureates photographed at the yearly ceremony, someone else may think of grey-haired nutty professors messing around with test-tubes. Complex and multifarious, based on fact and fiction, these images of scientists are constantly recycled and revised through various media. The reciprocity between cultural beliefs about science and the popular representation of the scientist has been traced back to the beginning of the twentieth century.[32] In the early 1900s, scientists were largely perceived as otherworldly wizards, operating outside the sphere of social life. Only after the invention of the atom bomb, when concern about their public image prompted scientists to become more involved in political action, they started to invest in their reputation as responsible citizens. In addition to their professional role, scientists were ascribed a social role and a moral responsibility. The balancing of scientific and social responsibilities is at the core of many public skirmishes over genetics.

Images of geneticists also inevitably elicit archetypal images of scientists; these stereotypes are not created in a single article or account, but no story is uninfluenced by the typical images locked away in our collective consciousness. Individual scientists are often attributed fixed features that have been perpetuated through literature, movies and magazines. They are not seldom depicted as obsessed maniacs, arrogant hermits or unwordly virtuosos.[33] Except for a number of general stereotypes, literature has endowed the public imagination with a number of specific impersonators of science, who are now referred to as archetypal models. Goethe's Faust, for instance, epitomizes the Romantic belief that preoccupation with science atrophies the emotions normally sustaining personal and social responsibilities. Stereotypes of scientists, benevolent or evil, are often invoked to confirm pre-existing anxieties or hopes.

Images of geneticists not only materialize through verbal evocation of mental concepts, but increasingly through visual illustrations and actual photographs.[34] Until the 1970s, pictures of scientists were still rare in popular stories. If photographed at all, scientists were usually represented as part of a group in the non-descript, passport-size pictures that illustrated the articles. In more recent popular narratives, however, the scientist is pushed into the limelight of public attention. Molecular biologists and geneticists, dressed in white lab coats and surrounded by sophisticated high-tech equipment, increasingly pose for magazine photographers. Photos of individual geneticists against a backdrop of 'their' instrumentation, are often paired off with colourful enlargements of microscopic images that make infinitessimal molecules visible to the ordinary eye. The popularity of the new genetics is undoubtedly enhanced by scientists' improved ability to visualize their objects of research with the help of technology. Photographic images function as irrefutable evidence – a valuable asset in the struggle to promote a particular definition of genetics. The visibility of scientists coincides with an ostentatious presence of their visualizing instruments. Through the lens of the researcher, readers are presented with the scientific point of view; after all, popular magazines often prefer visually attractive narratives over complex stories with various perspectives.

The transformation of genetics proliferates via a constant alternation of generic and innovative images. Just as the image of the geneticist can be manipulated by using strategic casting techniques, the meaning of genetics can be managed through the narrative device of plotting. A structure that seemingly connects and records events, a plot informs a specific interpretation. It is extremely difficult to explain to a general audience what geneticists do or what their work entails. Popular science stories usually present scientific 'events' as a teleological structure: experiments resulting in a discovery. In gene narratives, the order of telling is a reconstruction of the logic of argument, never an account of experimentation.[35] Apart from a sequential order, a popular gene narrative also requires suspense; without suspense, science is nothing but a dull sequence of experiments. Several types of plot typically structure popular gene narratives; the framing of genetics as a 'race' probably tops the list of gene-discovery books published in the last ten years, followed by the plotting of genetics research as a detective or an adventure story.

Plots or narrative frames impart much more than just the sequence of events that lead to a discovery: they also inscribe symbolic meaning to the subject and the actors. A recent BBC/PBS production entitled *The Gene Race* may illustrate how the use of multiple story-lines amplifies the dramatic impact of 'gene discoveries'.[36] The documentary's opening shot stages a car racetrack. When one of the race drivers gets out of the car, he turns out to be a nineteen-year-old British boy who suffers from cystic fibrosis (CF), a hereditary lung disease. The second story-line follows the boy into the hospital where he is engaged with his doctors in a 'race against death'. The average child with CF will live until the age of twenty-seven, and doctors have embarked on an exciting experiment with gene therapy to cure the boy's affliction. Finally, the plot is mobilized to frame the research process as a 'race' between a British and an American team, who compete to find the first successful gene therapy treatment for this rare disease. Besides imposing a sequential ordering onto a scientific 'event', the race-plot relates obvious symbolic connotations. The race between research teams implies that competition in the world of science is as healthy as competition in the world of business – a capitalist trope for technological prowess and progress. The race against death adds an aura of heroism and sainthood to geneticists. And the literal car race equates genetics to a sports game, a popular and entertaining pastime that draws large crowds. Identification and interpretation of single or multi-layered plots reveals the various connotative levels of meaning production.

Not only are plots employed to sculpt the image of genetics, but this image is also moulded by casual allusions to well-known literary fictions or myths. Invocation of familiar plots may help associate a scientific discovery with historical or cultural highlights. By far the most popular fictional source drawn upon for this purpose is *Frankenstein*. In the process of reframing an old story to sustain the plotting of a new one, the inherent ambiguity of the original story often gets lost. The rich content of Mary Shelley's *Frankenstein or the Modern Prometheus* is often reduced to an experiment out of control caused by an obsessed scientist who tries to penetrate Nature's secret.[37] Yet anyone who reads Shelley's novel carefully will be surprised at how unwarranted this commonsense interpretation really is. *Frankenstein* describes as much the failure of humanity as the failure of science. If the scientist is to blame at all, it is not for creating and using his monstrous technology, but for his refusal to take any responsibility for the consequences of his act. Many literary critics have pointed out how Shelley's story does nothing to support the reductionist contention that Frankenstein impersonates the evilness of science, but rather reflects upon the callousness of humanity in relation to their scientific inventions.[38]

Almost as popular as Frankenstein as a narrative source for popular genetic stories is Aldous Huxley's *Brave New World*.[39] Huxley's tale of standardized mass production of human beings, published in 1932, was as much a social critique of a hierarchical society than a critique of science.[40] Like *Frankenstein*, *Brave New World* is often invoked to underline contemporary anxieties of technology threatening to dehumanize humankind. Rather than fixing the locus of power abuse solely in science, *Brave New World* also points a finger at society. The novel instantiates the logical development of a consumer- and technology-oriented society whose predominant tools are scientific and whose goals are self-perpetuation and material progress. Peculiarly, this critique of capitalist culture rarely resurfaces in recent retellings of *Brave New World*. The invocation of Huxley's morality play appears to refer uniquely to his alarming image of artificially produced standardized individuals. Contemporary stories on cloning invoke Huxley's imagery of assembly-line-produced Alphas and Gammas without the vestigial anti-materialist critique directed at the ruling classes of a society, who maintain the lower classes happily consuming and producing. In its popular reknittings, *Brave New World* is similar to *Frankenstein* in that an important critical dimension of the original story gets lost in the act of reframing.

Due to its self-acclaimed concern with the 'secret of life', gene narratives are also frequently embroidered with biblical or mythical plots. The book of Genesis is often recited in popular genetics stories, even if these stories lack any further reference to religious exertions. Religious origin stories tend to reinforce the high moral grounds that the new genetics claims in its quest for the 'essence of human life'. Origin stories may have both a sacral and secular tenor, as evidenced by the story of the Holy Grail, a recurring source in popular stories on the Human Genome Project. The invocation of conventional frames and the incorporation of myths considerably affects the plotting of popular gene stories. Popular retellings often yield a univocal interpretation of the original story that projects a prefixed meaning onto genetics; consequentially, new stories and myths are (re-)created through the interpretation of old ones. *Frankenstein*, *Brave New World* and other tales emerge so frequently in contemporary tales that it is almost impossible to understand the popular image of genetics without a profound knowledge of these shared cultural narratives.

Just as characters, plots are not exclusively deployed to attach 'fixed' images of genetics onto new technologies or scientific developments. The very same plottings may be reinvented to propose alternative interpretations. Well-known and worn-out plots can provide an attractive basis for the subversion of hegemonic meanings. Since the audience is familiar with commonsense interpretations, it can relate both to conventional and innovative plot structures. For instance, to present genetics not as a race but as a democratic process, or to interpret Frankenstein's enterprise not as the immanent failure of men playing God but as the failure of men to keep track of their technological innovations, may substantially affect the representation and perception of genetics.

While images of geneticists commonly take shape in the form of characters, and images of genetics can be retraced in the various plots, the image of the gene, then, manifests itself primarily through the use of metaphors and models.[41] The concept of a gene or a molecular structure is a very difficult concept to convey, and for its public understanding scientists have to rely on representational analogies. Models and metaphors provide a recourse to analogy when words fail; they resemble each other in the sense that both introduce new meanings, and transfer a construct from one domain to another. Both models and metaphors are hypothetical propositions of reality, and are inherently slanted and incomplete as modes of representation.[42] The difference between models and metaphors, however, is that while scientific models are based on verification and logical consistency, metaphors are often intentionally imperfect. By far the most famous model in science is the double helix, the double spiralling ladder which was proposed in 1953 to elucidate the structure of DNA; the best-known metaphor for the gene is probably 'code'.

Metaphors are crucial narrative tools in the popularization of knowledge; they provide prototypes for imaginary creations, serving to conceptualize something that is as yet non-existent, that is too abstract, too tiny or too large to describe in ordinary language; they offer new 'ways of seeing'.[43] To compare a gene to a video tape and the genome to a video recorder, for instance, consciously transfers the everyday language of video technology onto a complex scientific concept, which at least gives a lay audience the illusion that they understand the basic mechanism of genetics. Metaphors, like characters and plots, are inundated with implied meanings; they have the potential to impose prefixed or innovative meanings onto scientific concepts. When a new metaphor is introduced, its figurative meaning is initially recognized as such, but gradually it may become so ingrained in everyday (scientific) language that it loses its metaphorical meaning, and comes to function as a figure of speech – a process of 'demetaphorization'.[44] We call those metaphors 'root metaphors': they arise out of common sense to become conceptual archetypes, just prior to becoming icons or myths.[45] The 'genetic code', for instance, is hardly recognized as a metaphor any more, but has come to serve as a literal term. The metaphor of the human genome as a 'map' of human's genetic make-up is well on its way to losing its figurative meaning.

Between the 1950s and the 1990s, wide-ranging metaphors have been used to explain the concept of 'gene' in popular stories. The gene has been reconfigured as a 'code', a 'factory', an 'escaping bug', an 'alphabet', a 'book' and a 'computer program'. Metaphorical images have taken turns in dominating the scene; where some have vanished, others have remained remarkably persistent over the years. Along with the alteration in preferred metaphors, however, comes a change in conceptualization. The gene as 'code' proposes a succinctly different interpretive framework from the gene as a 'computer program'. The transition from the context of communication to computer science may seem like a simple extension of a con-

ceptual domain, but it has huge philosophical and ideological consequences. Metaphors work heuristically, as they mutually affect the
domains from which images originate and the domains onto which they
are transferred. The analogy between video technology and genetics not
only turns the 'natural' mechanism of genetics into an ostensibly simple
technological concept, but it concomitantly affects the image of video
technology, rendering it more 'biological'. Metaphors are launched in the
public domain to (re)inscribe a particular interpretation of what a gene is
and how it functions. While the emergence of certain metaphors, such as a
computer program, can be obviously linked to general technological or
historical developments, the preference for certain metaphors never
evolves 'naturally'. The choice of metaphors is always strategic, and scientists, journalists and others have always been keenly aware of the impact
of metaphors on the general understanding of science.

Characters, plots and metaphors are the most prominent narrative instruments in the 'performance' of the genetics theatre. They can be used to
(sub)consciously influence the perception of a scientific concept, a practice
or its practitioners. Images and imaginations are translations, they add particular unarticulated or implied meanings to explicit statements or arguments. Popular stories of genetics are all structured by these common
narrative techniques, which help create, sustain and modify hegemonic
meanings. Since gene narratives inform and reconstruct our collective
imagination, they constitute the core of genetics' transformation.

GENETICS AS STAGE PRODUCTION

Beyond the level of performance – a public play in which images are
struck and contested by social groups – we can also look at the theatre of
genetics from the perspective of its production. The necessary activities to
put up a stage production are usually executed by a number of professional
groups. Script writers, producers, reviewers and public relations managers
all have their succinct tasks in the production and promotion of a play.
Looking at the theatre of genetics as an 'image-producing process', it
makes sense to distinguish between the activities of *scripting*, *staging* and
setting.

Popular stories of genetics are often modelled on the basis of *scripts*, a
term that may refer both to the written text of a play and to a pre-inscribed
use of an object or the position of a person. Every actor or character is
given a role on the basis of a script, such as a protagonist or an antagonist.

Translated to genetics, most actors are cast in one role or another as they seem to hold fixed, unambiguous positions 'for' or 'against' genetics. Scientists are typically resisted by non-scientist – moralists, ethicists, feminists or activists who reject the claims of scientific progress on other than scientific grounds. This bipolar distinction between scientists and Luddites, however outdated and distorted, is still frequently used to classify people involved in, or concerned with, genetic engineering. Scripts of popular stories usually provide for unambiguous roles on the basis of professional typologies: scientists are prefigured as pursuers of disinterested knowledge, while non-scientists opposing science are invariably seen as having a 'special' interest, or holding a 'partial' or ideological position.

The pre-inscription of roles is particularly evident in the gendered nature of the characters' positions. By virtue of their ascribed feminine characteristics, women hardly fit the category of 'scientist' at all; they are allotted minor roles, and their representation qualitatively differs from their male counterparts. A special oppositional role is reserved for feminists, who counter so-called neutral scientists by posing as politicized anti-scientists. Yet while some groups of feminists explicitly endorse this apportioned variant of the classic Luddite position, others fiercely reject any prescribed role. The 'feminist', much like the 'scientist', is equally subjected to the vagaries of narrative scripting.

A perceived or preconceived division of roles is also evident at the level of *staging*. Various professional groups are involved in a staged production, as they are trying to draw public attention to the play. Whereas actors are supposed to concentrate on the acting, public relations managers take care of the play's promotion, and journalists inform the audience of its content, while also providing an evaluation. In the theatre of genetics, scientists are supposed to be the actors who have no power over the public evaluation of their act. Journalists are held responsible for the creation of popular meanings, whether in the form of favourable or 'distorting' reviews. Yet these pre-inscribed roles are not defined once and for all; since the 1950s, the dynamics between various professional groups mediating the popularization of genetics have considerably shifted.

Although scientists generally look down upon popularizing activities as inferior and distracting, they have long acknowledged the importance of popularization. But especially since, in the late 1960s, science funding became increasingly dependent on its reputation with government agencies and corporate sponsors, scientists' attitudes towards the public's perception of their work have drastically changed. They began to open up more to the press, and took an effort to translate profuse theory into under-standable, accessible prose. For a long time after World War II, journalists uncritically adopted scientists' explanations of their work, thus yielding them power to set the terms for public debate.[46] At the same time, however, journalists also took pains to impersonate the public's hopes, fears and anxieties. In tune with a general, post-Watergate shift in profes-sional attitude, science reporters became more adversarial, investigative and interpretive. This shift in vocational duties had repercussions on the relationship between scientists and journalists, and on the way they delineated their professional responsibilities.[47]

A significant factor in the changing dynamics between professional groups was the advent of public relations officials, who increasingly came to moderate the exchange of information between scientists, journalists and the audience. Science, like any company, increasingly had to draw attention to its products, which needed to be sold to investors, politicians and the public at large. Public relations strategists are now common in research departments or research and development offices of large biotechnology companies. They are actively involved in the marketing and promotion of a specific product or an entire organization – commercial or non-profit. Although scientists often claim to have no influence on the public perception of their activities, public relations 'packaging' has become an intricate part of product development. Public acclaim is no longer taken for granted, as scientists understand the big stakes in the popularity game. For journalists, a growing presence of public relations officials has profoundly affected their routines and practices. They find themselves more and more overflowing with prepackaged images and interpretations of science – ready-made nuggets of information.

Recently, we can notice an increasing recognition of popular culture – (science) fiction, television, documentary and film – as a contributing factor to the public recognition of a scientific operation or of scientific ideas. Traditionally, products of popular culture have been located outside the realm of 'serious' public debate, to distinguish it from factual genres such as (science) journalism. Because an ever-increasing portion of popular science is distributed through channels of popular culture, agen-cies have started to sponsor and support the production of movies, docu-mentaries and television series. The Human Genome Project, for instance, has been the subject of numerous documentaries and television series, in part paid for by the Project's co-ordinating office. Documentary or dra-matic narratives lend themselves particularly to the explanation of abstract goals of, and needs for, scientific projects. In these productions, scientific facts and imaginary tales are often intertwined, as producers rely on the devices of story-telling to structure their products. Despite their distinct roles, scientists, journalists and fiction writers often deliberately contest each other's discursive terrain, deploy each other's strategies and use the other's resources. An analysis of genetics as a theatre production allows to recognize such dynamics as a constitutive element in the image-making process.

Performances or plays are always staged in a specific *setting*. The setting of a theatre may be the physical location of the performance, but may also be a symbolic locale – a décor or background. While the most obvious physical setting for the theatre of genetics is the laboratory, genetic engineering also takes place in industrial environments and the hospital. The choice of physical settings determines which aspects of genetics can be admitted to a discussion and which simply do not apply. In a hospital, for instance, we expect different questions to be raised from those in a laboratory – questions concerning cures and patients. In addition to a physical place, theatrical setting may suggest symbolic space. The configuration of genetics often takes shape through the use of spatial analogies. Perhaps counterintuitively, the spatial dimensions of the DNA-molecule may be inscribed in terms of the cosmos; geneticists are repeatedly referred to as 'astronauts' of the new science, and genetics as a space adventure. On a symbolic level, it is more heroic to be a small person facing a big challenge than to be a person mastering a phenomenon that is too small for the eye to see. Changing the spatial dimensions of a cell into a universe equals the shifting of camera positions: it drastically alters the viewer's perception of an endeavour.

But spatial analogies may also attach a negative image to science. The story of genetic engineers tinkering with life's essence is often set in an 'underground' lab or space, connoting the evil or corrupt forces of a science that conspires against humanity. Whether we go 'up in space' or 'go down into' a subcellular realm makes an enormous difference; orientational metaphors attach a specific emotion or value to directional markers.[48] Historically, great scientific discoveries are associated with widening the horizon rather than closing the doors of the laboratory or office. Spatial metaphors and analogies contain bits of narrative without supplying the diegesis as a whole. Like images in general, symbolic settings allude to, rather than explicate, interpretations of science.

An analysis of the production of the theatre of genetics – its scripting, staging and setting – in conjunction with an analysis of its staged performance – its characters, plots and metaphors – should give us insight in the construction of popular images and imaginations. In addition to these two layers of analysis, the theatre concept allows for another entrance for assessing the mechanism of science popularization.

IMAGENATION'S CULTURAL CONTEXT

Like a theatre performance or production, the definition and implementation of a scientific theory or practice is tied up with historical, social and cultural changes. Although the public's recognition and validation of certain scientific paradigms always takes place against a backdrop of contemporary historical developments, the transformation of genetics is more than the consequence of a fluctuating *Zeitgeist*. Between the 1950s and 1990s, popular meanings of genetics have been continuously regauged in the wake of changing national and global politics, from Cold War tension to international détente. Changes in meaning are also reflections of shifting social concerns; genetics forms a continuous locus of contestation between capital, labour and power. Mounting environmental concerns in the 1970s evidently had a great impact on the public discussion of DNA-technology. And the exponential growth of biotechnology in the 1980s can only be properly understood in the context of deregulation and economic *laissez-faire*. Hence, popularization of genetics should be accounted for in the context of larger *historical* and *social* changes. The transformation of popular images, however, draws particular attention to the *cultural subtext* through which they proliferate.

The inscription of cultural change may be less obvious, but is not less significant than social or historical influences on scientific developments. People interpret ideas, images, stories and points of view in part by melding scientific representations with cultural styles and trends. Popular representations of genetics can be seen as an object form of cultural criticism. From the very onset, the debate over the theories and implications of genetics has given rise to philosophical concerns about the potential transformation of society, nature and culture. How should science be organized in society? How does genetics affect our view of culture? Can genetic engineering lead to a retooling of the natural body, and how does it redefine that body in the first place? This kind of question invites imaginary reconceptualizations of both technology and society, nature and culture. Traditionally, such critical re-imaginations are supposed to be contained within the domains of fiction and art – the assigned discursive spaces for unrestrained experiments with imaginary tools. As genetics evolved from a primarily scientific theory into a preferred way of thinking about health and disease, its function as a locus for cultural critique expanded from the discourses of art and fiction into a variety of other discourses that scaffold the dissemination of genetic knowledge.

The premise of the 'geneticization' of society does not necessarily mean that genetics has become more 'popular', but that its axioms and principles have spread out through an ever-growing number of discourses. This gradual incorporation of scientific knowledge into other than scientific domains can be understood as part of a larger, cultural trend that is generally referred to as the transition from modernism to postmodernism. The popularization of genetics should be tied in with the transformation of culture, just as it has been tied in with historical and social trends. Without entering the lengthy debate on postmodernism and its discontents, some insights into the transformation of culture may help understand the construction of popular genetic knowledge. Postmodernism has invariably been defined in opposition to modernism, and two major 'shifts' that have been identified as typical of this transformation may be relevant to our discussion on the popularization of genetics. In light of an emerging postmodernist culture, theorists have pointed at the mounting emphasis on 'images' rather than 'content' of science. The victory of the simulacrum over the referent – or the emergence of the ontological dominant at the expense of the epistemological dominant – has been extensively decried and theorized.[49] Second, it is commonly assumed that, in the postmodern era, science and philosophy have jettisoned their grandiose claims to truth, and have reappeared splintered via a number of discourses.[50] The demise of 'grand narratives' – all-encompassing scientific paradigms and universal social theories – has resulted, according to philosophers, in a complete 'narrativation of knowledge'. Although the hegemony of the image-without-referent and the narrativation of scientific knowledge by now qualify as truisms in the characterization of postmodern culture, how does the identification of this trend help us understand science as an image-making process?

Instead of accepting 'modernism' and 'postmodernism' as self-evident, explanatory categories, I will use the generally assumed vector of transformation as a mode of inquiry to investigate manifest tensions between 'fixed' and innovative images of genetics. This cultural subtext helps raising questions about persistencies and innovations in the imagination, about the 'fixedness' and the manipulability of images, and about the permeable borders between science and culture. By systematically rethinking cultural concepts, one can tie in discursive change with larger social or political practices. For instance, how and to what extent did the emergence of sociobiological discourse coincide with capitalist incentives that propelled the business of biotechnology? Why do certain worn-out images of genes, such as the book or language, re-emerge in the fully digitilized practice of human genome research? And how is the discourse of genomics injected with a cultural politics that inscribes the standardization of the human body? The transition to postmodernism, as a cultural subtext, will serve to identify, rather than solve, apparent contradictions and tensions.

GENETICS AS SPECTACLE

The conventional 'arrow of progress' metaphor espouses a view of science as a process where theory develops into technology, technology is turned into industry, and market changes in turn lead to social progress. Analogously, the classical 'diffusion' model accounts for the dissemination of scientific knowledge as a linear flow of information moving top down, from scientists via journalists to the general audience. As an antidote to these dominant views, I propose the metaphor of theatre to scrutinize 'geneticization' or 'popularization' of genetics. Rather than providing a theory or model, the theatre metaphor serves as a prism for looking at 'geneticization' in terms of circularity instead of linearity. 'Theatre' provides a critical, self-reflexive tool for figuring out what is being said by first figuring out who is saying what to whom and for what reasons. It offers a multi-layered viewpoint that accounts for both the production and (re)presentation of science in the public domain, while acknowledging its intrinsic embeddedness in society and culture. Geneticization, in this line of reasoning, is not the gradual inculcation of genetic ideas in the public mind, but a gradual expansion of loci of contestation where meanings of genetics are weighed. Genetic knowledge is not fabricated by scientists and obstructed or promoted by social groups, and neither is it mediated favourably or unfavouraby by professional groups. The dynamics between the social and professional groups are part and parcel of meaning production. To treat genetics as a theatre is not to be dismissive; on the contrary, the metaphor simply assumes that the natural sciences are as culturally complex as the human sciences and the arts.

The theatre of genetics is defined by its story-tellers: they endow entities with qualities, they classify actors and attribute myths to science and nature. My aim is not to 'go along' with the gradual geneticization of culture, but to recognize its many contradictions, point at the subversions,

Imagenation

and illuminate the places where hegemonic and subversive images clash. This analytical process is concomitantly an archaeological process: I follow the interplay and exchanges of images and imaginations, to trace how knowledge is disseminated through a variety of domains.[51] Genetics cannot be reduced to a single script, play or performance; it should rather be seen as a successive flow of performances enacted on a permanent stage, where actors continuously walk on and off. The theatre is a place where (inter)action is regulated by discursive conventions and practices. Hence, narrative analysis does not only apply to the performance on stage but also to its production and context.

I will trace the transformation of genetics as a play in four 'scenes'. Each scene signifies a moment in the story of geneticization where a significant shift in popular representation takes place, and where new claims or practices lead to dissent or dispute. Popularization, after all, can best be studied at moments of controversy or absence of consensus.[52] The first scene relates the introduction of the 'new biology' in the 1950s and 1960s. Scientists and theologians were the prime contenders in this stage of public negotiation, as they shaped the images of genetics in terms of morality. For the second scene, we move to the fierce DNA-debate in the 1970s, when genetics became the focus of a political controversy over biohazards. Environmentalists and feminists entered the discussion on DNA by contesting scientists' power over the interpretation of scientific knowledge. The third obvious moment for studying the popular imagenation is the growth of the biotechnology business in the 1980s. Scientist-entrepreneurs and sociobiologists crossed swords with anti-capitalists and feminists in a public contest over the evaluation of a biotechnology boom. Finally, the initiation and implementation of the Human Genome Project provides another moment for dissecting the contestation of popular images. A number of social and professional groups are involved in the promotion of this giant project, but the 'biophoria' of genome mapping is countered by a variety of detractors.

These four scenes do not represent chronological stages in the public evaluation of genetics, but are retrospectively constructed markers that delineate the changing contexts in which this science is publicly communicated. Naturally, each of these 'stages' or 'scenes' evolved at a particular moment in time, reflecting historically changing political and social concerns. Although the linear order may impart the notion of a strict chronological sequence of events, these scenes or stages rather signify epistemological and conceptual shifts. In fact, the ambiguity of the word 'stage' can be taken very literally: 'stage' concurrently refers to a point or a period in a process of development, and to a raised floor or platform where acting takes place. There are no 'great breaks' in this transformation, and its description is a description of a heterogeneous exchange of symbolic values. Evaluative frameworks appear conjuncturally rather than successively, and always partially overlap. While in general we can designate a turnabout in the public appreciation of genetics in the late 1970s, this does not mean there has been a steadily growing, univocal acclamation. Instead, each 'stage' can be characterized as a clash between hegemonic and critical evaluations of genetics.

Opting for the metaphor of theatre, I deliberately shift attention from the social to the cultural dimensions of science. The image of the world of science – the image that the public perceives – is not that of 'science in action'. Science-in-action is the science as it is performed off-stage and in the wings. Evidently, science-in-action is also a sort of performance, sustained by rhetoric and ritual, as social constructivists have convincingly shown.[53] Science-in-action is never completely divorced from 'science as public performance', as they involve some of the same players. Yet a cultural analysis of genetics will focus on what happens on stage, in the public spotlights that illuminate the drama enacted by scientists and all other groups engaged in the popular representation of genetics. The theatre offers a microsociological unit of analysis, and narrative analysis furnishes a microscope through which the mechanisms of image-making and imagination are magnified.

We know from the history of drama that some plays have drawn huge crowds, and, as a result of their popularity, have been staged over and over again. The audience is both stake and stakeholder in this performance: the whole purpose is to have as many people as possible 'buy into' a particular meaning of genetics, hence increasing public support. We do not know what causes some plays to be more successful than others. Some plays gain recognition through their famed scripts, others because of their actors' brilliant performances, again others have attracted large crowds after glowing reviews in newspapers. The popularity of a play, as everyone knows, can never be predicted by reducing success to a simple formula or 'code'. My reconstruction of genetics' popular images is not meant to construct a uniform model for popularization, but intends to provide a tool to analyse the numerous variations of genetics' popular manifestations. This analysis may cast light on the role that images and imagination play in the dissemination of scientific knowledge, in conjunction with the changing dynamics of 'image-makers', and in the context of changing historical and cultural conditions. The endless varieties coming out of that process, rather than a standardizing model, form the focus of this inquiry. *Imagenation* is an attempt to reconstruct the stories of genetics, assembling in retrospect the multiplicity of facts, acts, factors and actors that have contributed to the 'geneticization of culture'. Genetics is a spectacle, and the purpose of this book is to enter the popular contestations and contribute to its continuously re-created 'imagenations'.

2 Biofears and Biofantasies

A NEW IMAGE FOR A NEW FIELD

Although the 'beginning' of genetics can be traced back to the turn of the century, when it became known as the science that deals with heredity and variation of organisms, in the public mind genetics commonly starts with the 'discovery' of the helical model in 1953. Since that particular moment, genetics has ostensibly developed as *the* (rather than *a*) molecular biology of all cells. Molecular biology, the discipline that is concerned with the physicochemical organization of living matter, had emerged in the late 1930s, but proliferated in the 1950s as the 'new biology' or the 'new genetics' – a discipline solely focusing on the molecular basis of inheritance. Yet despite the retrospectively claimed grand turning-point, genetics did not start a triumphant offensive in 1953. For one thing, it took until the late 1960s before a satisfactory theory had been formulated on how genetic information contained in DNA governed the making of proteins in organisms. Moreover, postwar genetics research was haunted by tainted images and compromised memories of formerly prevalent eugenic theories. The fear for abuse of genetic knowledge, in the shadow of Nazi atrocities and anxieties held vivid by lingering Cold War tensions, cast a dubious light on a science that embroidered on the same themes and theories as eugenics.

In emphasizing the 'newness' of their field, geneticists and molecular biologists wanted to shake off ideologically compromised connotations, and establish their field as socially useful and scientifically respectable. The 'new genetics' claimed an entirely new set of goals and potential applications, thus trying to distance itself from its tainted precursor. Nevertheless, the new goals were publicly evaluated in the context of past implementations of eugenic theories. Geneticists had to negotiate their claims to a new beginning within the moral parameters of the public's perception of genetics. During this first 'scene' in the theatre of genetics, scientists were ostentatiously counteracted by special interest groups like moralists or clergy, and by professional groups like journalists and fiction writers. Both were regarded as hindrances on the way to public acceptance. Yet how did geneticists actually attempt to create a 'new' image, and how were scientists 'resisted' by various groups in the dissemination of knowledge?

Changing the unfavourable public opinion of this scientific discipline was no small feat. Since the dropping of atom bombs on Japan, the moral stocks of scientists in general had plunged, and ongoing experiments with nuclear bombs in the USA only added to public uneasiness with science. In the wake of the Manhattan Project, new scientific projects were met with anxiety and distrust. Scientific experiments were mentally linked up with the nuclear age and nuclear physics seemed to be the discipline most visible to the public eye, due to its immediate military relevance. The need for nuclear physics, at the height of the Cold War, seemed obvious to a general audience: nuclear bombs were a necessary evil, the only way to keep 'our enemies' in place. The need for the 'new biology', though, was less self-evident. It was not exactly clear what other than dubious uses the prospect of 'germinal choice' offered to society.

Until the mid-1960s, media coverage of genetics was sparse; the few articles on genetics that we find in popular magazines and other sources reflect ambitious attempts to establish a new image for a new field. The urge for the 'new biology' materializes against the backdrop of two contemporary social concerns: the need to find a scientific remedy to the harmful effects of nuclear radiation, and the need to offer a possible solution to a postwar population explosion. Recognition of these needs seemed a precondition for changing the public's esteem for molecular biology and geneticists. The articulation of new needs rarely happens explicitly; rather than through plain arguments, the new needs for genetics proliferate through association, with other established disciplines, away from eugenics, and through the launching of new metaphors for the gene.

The idea that there is an urgent demand for the new genetics takes root via initial association of genetics with nuclear physics. After the bombings on Hiroshima and Nagasaki, American experts had been sent to Japan to examine types of cancer caused by the release of unprecedented amounts of radiation. In the late 1950s, experiments with nuclear explosions in the Nevada desert were surrounded by secrecy and the dangers of radiation were effectively downplayed by government agencies. Geneticists, as we can read in several news magazines, were primarily concerned with finding a potential cure for the damaging effects of cell mutation. Not only may genetics research offset the harmful effects of nuclear radiation, but it aspires to the grander aim of finding the essence of life. Geneticists 'do not dodge questions about the origin of life on earth' as *Time* magazine boasted in 1958.[1] The comparison with nuclear physics positively set genetics as the 'study of life' alongside the destructive forces of the atomic bomb, signifying the 'secrets of death'.[2] Public suspicion of high-tech as a war-time villain seems the flip side of the postwar adoration of science as the new salvation; genetics is constructed as an essential weapon in the Cold War political paranoia.

While genetics was presented as an antidote to the devastating consequences of nuclear physics, it was nonetheless associated with a discipline that had a solid academic status. According to the same *Time* article, the 'comparatively young, fast-developing science of heredity' aims at solving the mystery of life just as 'physics works at solving the mystery of matter'. The evil effects of nuclear physics can be outweighed by the beneficial effects of genetics, another 'young science'. Molecular biology may actually prevent excessive release of nuclear material in future explosions, argues *Time*: 'The comparative cleanness (low fallout) of the test bombs that the U.S. was exploding last week [in Nevada] was in large part a response to the warnings of the geneticists' (50). The report suggests that deeper knowledge of the inner structure of human cells may demonstrate the harmful effects of nuclear radiation on cell growth. Without explicitly degrading the work of nuclear physicists, the journalist ascribes to genetics the same authority as nuclear physics, while at the same time disapproving of the latter's applications.[3]

Through its association with an already established discipline, the new biology also tries to undo itself from its ideologically compromised past. Ambitious new goals sharply contrast previous goals of eugenics to optimize the population's genetic 'quality'. The fact that allusions to Nazi practices form the unarticulated subtext for the definition of the new discipline can be derived from a categorical denial of any continuity between the old eugenics and the new genetics. Though the scientific roots of molecular biology are respectfully acknowledged – the work of Mendel, De Vries, Beadle, Morgan and Muller pass in review – their work is never explicitly related to eugenic theories widely accepted in the USA and Europe in the 1920s and early 1930s. What we get from this 1958 article, is a respectful enumeration of virtuous scientists who have paved the route for a promising new field. The idea that the 'new biology' indeed marks a new beginning is further amplified in a 1963 cover story in *Newsweek*.[4] The 'new genetics' no longer starts with Mendel, but now begins with the 'discovery' of DNA by Watson and Crick, whose 'masterly piece of analysis and imagination' revolutionized genetics, and will undoubtedly help scientists to find more 'secrets of life' (63). No predecessors are mentioned, and Watson and Crick are hailed as the intellectual fathers of a revolutionary biology. Judging from this issue of *Newsweek*, genetics had been 'invented' ten years earlier, in 1953, so the new genetics is rhetorically cleansed from its tainted history.

A second way in which the new genetics is disconnected from its compromised roots is through the introduction of gene images. Hardly any popular stories on genetics include an explanation of scientific details, but the *Newsweek* reporter tries to educate an ignorant lay audience with the help of a few telling metaphors: the 'alphabet', 'language' and 'genetic code'. The four bases of DNA – A (adenine), G (guanine), T (thymidine) and C (cytosine) – are labelled as the letters in the alphabet of DNA, and the endless combinations of letters constitute 'sentences' in a 'book'. All the letters of our DNA combined form the instructions for our genetic make-up, and knowledge of this structure may lead to the control of heredity. Language-related metaphors are used interchangeably with 'code', 'message' and 'inscription'. If applied to the structure of DNA, 'code' infers the idea of a rule-governed system of communication that can be understood by everyone who has a key to its formative principles.

Neither the language nor the code metaphor were exactly new in 1963. The idea of the 'legibility of nature' or 'the book of nature' is a common trope in the life sciences.[5] Erwin Schrödinger is believed to be the first to have coined the term 'code' to describe the principles of heredity in 1944.[6] Although both metaphors had been introduced at an earlier stage, they were rendered more poignant in the context of the double stranded helix, as they came to signify the 'coded key script' to life's molecular essence. The genetic 'language' and 'code' signalled a definite shift from metaphors of the phenotype to metaphors of the genotype.[7] Out of the 'code' metaphor, an extensive web of related semantic strings evolved, often conflating artificial sign systems with everyday language. The metaphorical idiom seriously caught on in the 1960s, culminating in what Edwin Chargaff dubbed the 'grammar of biology'.[8] The extended 'code' metaphor by far exceeded the function of exemplification proper; eventually, it fed a new dogmatic biology based on the assumption that 'life' can be explained by, and reduced to, its molecular structure. The idea of an 'underlying code' enabled scientists to think of heredity as a general pattern that, once it is known, can be used as an explanatory model for countless phenotypical variations.

It may not be a coincidence that the 'code' metaphor gained wide acceptance in the 1960s, the decade in which Noam Chomsky's transformative generative grammar revolutionized linguistics, and Marshall McLuhan mesmerized the world with magical mantras involving the medium and the message. Metaphors, after all, change in conjunction with dominant theoretical and conceptual fashions. It has even been argued that linguistics, in the 1960s, came to the rescue of biology, offering it a structured analysis.[9] Genetics shared with transformational generative grammar the assumption that there is an abstract set of rules that govern all 'natural' linguistic variations. In accordance with McLuhan's theory, molecular biology espoused a deterministic vision of the medium that controls the message. For a long time, molecular biology, linguistics and communication theory shared a single conceptual framework via the concepts of code, message and medium. The terms 'language' and 'code' were obviously more than helpful aids in explaining the working of DNA. Beyond their obvious meanings, the terms add important connotations. 'Code' and 'language' transfer the ability to communicate – a profoundly human(ist) activity that lies at the heart of interpersonal and social relationships – to a molecular structure. Transposed onto genetics, it makes DNA appear as a lucid communication system. 'DNA as code' conjures up images of a sign system like Morse, which is highly transparent and which encoding and decoding rules are perfectly unambiguous. A preference for these kinds of linguistic metaphor derive directly from war-time obsessions with code breaking and defeating the enemy with Information Science.

Probably the most important, though least noticed, consequence of the widespread distribution of the code metaphor was that it turned the gene into a palpable entity, the foremost object of molecular research. Both the helical model and the gene metaphors enabled the representation of the human body in terms of its tiniest elements, analogous to the atom in nuclear physics. Like letters, genes were supposed to constitute the elementary units of living matter. The gene became reified as the central locus of vital activity, and molecules other than DNA were from now on ranked as secondary.[10] Its new verbal coating redefined the role of the geneticist as someone who is supposed to find the grand scheme of rules by which molecules are organized and transformed. Coding inevitably leads to ordering, inferring the notion that genetic mutation can be thoroughly systematized and decoded.

In these early years of molecular biology, another important metaphor is propelled into popular discourse besides 'language' and 'code'. The article in *Newsweek* compares the production of protein to 'an auto assembly line, where DNA, like a foreman, remains aloof from the actual labour and appears to send messengers out into the cell bearing precise instructions'.[11] The 'factory' is far from original as a metaphor to elucidate biological processes, but has traditionally been employed to exemplify a number of organic mechanisms, such as the cell and the reproductive system.[12] It allows us to view the genetic structure as a mechanical system which can generate products, but also as a system that can fail because at some point the assembly line may break down. Yet the gene is explicitly presented as a foreman, someone in charge of assembling processes, signalling that the gene's status is higher than that of cells or organisms. This 'natural structure' of DNA also seems to be held up as a model for class capitalism, in which the 'aloof foreman' justifies intrinsic inequalities between workers and supervisors.

The introduction of new images, besides illustrating the working of the new genetics, further amplified the ideological rupture between eugenics and the new genetics. Metaphors like 'language', 'instructions' and 'factories' relocate genetics from the ideologically infested realm of population policies to the apparent politically neutral context of the laboratory or the factory. As we can read in *Newsweek*:

> Despite the controversial background of eugenics, it would be undoubtedly merciful to *remove* the *instructions* for some of the 250,000 genetic defects in babies born each year. Indeed, *alteration of genes* in the *laboratory* may prove to be more acceptable than controlled breeding of humans; it certainly is preferable to sterilization of genetic misfits (66, emphases added).

Genetics is once again rhetorically separated from its tainted precursor eugenics, and these new metaphors also allow strategic distancing from doubtful alliances with nuclear physics.

In another part of the article, the *Newsweek* reporter uses the words 'genetics explosion' to refer to the rapidly increasing number of scientists from a variety of disciplines who are becoming engaged in genetics. While the word 'explosion' may still evoke reminiscences of the atom bomb, the term also alludes to another threat now frequently surfacing in the headlines of magazines: the fear for a man-induced, accelerated and uncontrollable expansion of the human race. An immanent 'population explosion' is repeatedly associated with the 'genetic revolution', implying that there is an urgent need for technologies that may help control and improve the quality of human life. This justification is never stated explicitly, but emerges as an unstated premise for the development of the new genetics.

THE DOUBLE HELIX

Negotiation of a new image for a new field did not exclusively occur through the establishment of gene metaphors, but also via the introduction of new plots and characters. The most significant attempt at upgrading the status of molecular biology and 'popularizing' the field of genetics was undoubtedly James Watson's *The Double Helix*. In his autobiographical account, Watson reminisces how he, with the help of a few other men, revolutionized molecular biology by discovering the helical structure of DNA.[13] *The Double Helix* became immensely popular after its publication in 1968, fifteen years after the actual proposition of the helical structure, and six years after Watson, Francis Crick and Maurice Wilkins collected their shared Nobel Prize. Watson's popular account was not really about the discovery of the helical model – the discovery was 'old news' in 1968 – but it was about claiming the discovery as the 'grand turning-point' or, perhaps more appropriate, the birth of a promising new science. Written by a distinguished scientist, *The Double Helix* has an easily accessible autobiographical format that gives the reader an uncanny insider's view of scientific practice. Very few press reports had ever paid attention to the actual experience of doing prize-winning science. Watson's prime aim in writing *The Double Helix*, besides propagandizing the technical superiority of their proclaimed method over classical forms of biological inquiry, was to raise the status of laboratory researchers in general and of geneticists in particular.

The most powerful tool in reframing genetics is Watson's use of narrative plotting. *The Double Helix* unfolds as a classic detective story: it relates how two men set out to find the solution to a theoretical problem and end up finding the 'secret of life'. However, a journey resulting in predictable triumph does not stir up the desired effect, unless a notion of suspense is added. Like most detective stories, the plot of *The Double Helix* revolves around a single protagonist, whose shrewd intuitions lead to the hidden treasure. At the very beginning of his account, Watson unashamedly presents himself as the only genius, sometimes aided by a few intelligent men. 'One could not be a successful scientist without realizing that, in contrast to the popular conception supported by newspapers and mothers of scientists, a goodly number of scientists are not only narrow-minded and dull, but also just stupid' (18). The young Watson is cast as a brilliant young man, an aimiable character who does not waste his time. As befits a predestined winner, he courageously defies bureaucratic decisions that interfere with his scientific goals and personal preferences. Arrogance and stubborness are character traits considered permissible in someone who is bound to be a prodigy; by the time of the book's publication in 1968, the now Nobel laureate can certainly afford to stage this kind of persona. The adventure plot underlying *The Double Helix* typifies the scientist as a hero, counterbalancing popular images of dull lab workers or suspect wizards.

The second plot that structures the autobiographical account is no less effective in adding notions of heroism. Watson is one of the first to present the search for a scientific paradigm as a *race*, as he frames the quest for the helical model as a competition. To that end, he recasts Linus Pauling's role somewhat awkwardly into that of a rival. He boasts how he secretly read through Pauling's letters to his son – who was working in the same lab as Watson at the time – thus monitoring his 'competitor's' progress in the field. Apart from Watson's retrospective account, there is little evidence that Pauling ever saw himself as Watson's rival. On the contrary, when Pauling visited Watson and Crick in Britain, he expressed his sincere admiration for their theoretical proposition.

The 'race' frame of Watson's story is doubly reinforced through Watson's retrospective casting of Rosalind Franklin as the main hindrance on his road to triumph. Together with Maurice Wilkins, Rosalind Franklin formed a 'rival team' apparently intent on beating Watson and Crick. Since Maurice Wilkins could not in retrospect serve as an obstacle – after all, he shared the Nobel Prize with Watson and Crick – Franklin becomes Watson's sole designated enemy. She is restylized to form the formidable obstruction the hero needs to underscore his titanic achievement. In the beginning of his book, Watson sympathizes with Wilkins who has to 'endure' the capriciousness and uncooperative behaviour of his female professional companion. Watson and Wilkins snigger at Franklin's firm belief in crystallography as an empirical method to find the structure of DNA. Later, Watson brags about how he talked Wilkins into giving him the X-ray photographs he stole from his colleague Rosalind Franklin. The X-rays admittedly formed the empirical basis of the helical model that he and Crick subsequently proposed.

The Double Helix has been passed off as a delightful digression on the way in which science is done, and has been severely criticized as a self-aggrandizing account of an ego-inflated scientist. Watson's attempt to change the common perception of molecular biology from a boring accumulation of lab tests into an exciting race, where teams of scientists compete with each other to solve the enigma of nature, has drawn serious criticism from various angles. Some scientists have interpreted the book as an attempt to redefine the mores of collaborative research. *The Double Helix* purportedly promoted a new professional ethic which encourages competition and justifies any means necessary to attain a set scientific goal. As Edward Yoxen argues:

All kinds of literary and psychological devices are drawn upon to defend a new realism about competition, aggression and self-confidence. This, Watson seems to be saying, is how science has to be done, if the fundamental questions in biology are to be laid bare and answered. Its function was to display a canonical example of a new scientific style, with ruthless predation on other fields and other's work, minimal courtesy to supporting colleagues and peers, continual defiance of troublesome data and a positive contempt for traditional intellectual concerns.[14]

While presenting his scientific model as an inscription of reality, Watson seemed to favour a scientific practice characterized by professional bickering and rivalry. Without a sign of bad conscience, he relates how he secretly cajoled people into giving him information.[15] Watson allegedly changed the rules of the game: rather than obeying the old generation's rules of courtesy, colleaguiality and openness, he promoted new methods such as wheedling, competition and secrecy. For this very reason, *The Double Helix* has been condemned as a distasteful proposition of how 'science is done'.

Not surprisingly, *The Double Helix* has also given rise to numerous feminist critiques, all pointing out the pervasive sexism in Watson's casting of Rosalind Franklin. Franklin's contribution to the model is not simply downplayed by Watson: he literally dismisses her active role in the research process. 'Rosy' – the diminutive with which Watson consistently refers to her – is a constant object of vile gossip. Besides derogatory comments on her appearance and clothing, his remarks about her character are downright mysogynist. Watson uses the word 'bluestocking' more than once to describe Franklin, and argues that the best place for a 'feminist like Rosy' is 'in another person's lab' (21). Anne Sayre has discussed Watson's hostile treatment of Franklin in the context of contemporary discrimination of women in science.[16] Comparing the Rosalind Franklin she reconstructs from personal memories and diary entries, Sayre concludes that Watson has constructed a fictional character to contrast his own success with the fruitless efforts of a weak and dumb female scientist. Hillary Rose has pointed at the gendered hierarchy of scientific methods displayed in the book: Watson's abstract 'masculine' method of model-building triumphs over Franklin's 'naive feminine belief' in empirical evidence – X-rays of molecules.[17] Indeed, Watson needed the proof provided by Franklin to arrive at the conclusions that won him his Nobel Prize. Had Franklin lived to 1968 – she died of cancer in 1957 – Watson's story could never have been published in this form. To accommodate some of the criticism his fellow scientists fired at him after reading the manuscript, he added a perfunctory epilogue stating that 'he realized too late the struggles that the intelligent woman faces to be accepted by a scientific world' (143). To many scientists' dismay – not least Crick's – Watson never retracted his initial portrayal of Rosalind Franklin.

The Double Helix seems, on the face of it, a case history of the pervasiveness of sexism in the production of science or the genderedness of scientific methods of inquiry. Similar to the criticism of Watson's story as an attempt to change the mores of scientific practice, some feminist critics have equally considered *The Double Helix* as a realistic account of 'how science is done'. Rather than a realistic story, however, *The Double Helix* could also be viewed as an ideological and symbolic inscription of gender and science. Rosalind Franklin becomes Watson's 'natural' enemy as the struggle for the discovery of DNA is modelled after the conventional battle between the sexes. Men and women assume oppositional roles in the scripted play: Franklin does not become Watson's main obstacle because she *is* a woman, but because she does not know *her place* as a woman.

If we compare Franklin with other female characters in *The Double Helix*, it becomes clear that Watson perceives women as creatures inherently ignorant of science. As illustrated by his description of Odile, Francis Crick's wife, teaching women anything scientific is a lost cause: 'At no moment did Francis see any point in trying to simplify the matter for Odile's benefit... Not only did she not know any science, but any attempt to put some in her head would be a losing fight against the years of her convent upbringing' (63). Despite their ignorance, however, women are not completely unuseful in the pursuit of scientific knowledge: they may serve as facilitators and communicators of scientific work. Watson reaccounts affectionately how his sister Elizabeth selflessly sacrificed a free Saturday afternoon to type the manuscript for *Nature* (140). Earlier on in the story, Watson details how he unsuccessfully tried to couple off his sister to Maurice Wilkins – a relationship which would have benefited his career, since 'if Maurice really liked my sister, it was inevitable that I would be closely associated with his X-ray work on DNA' (28).

The Double Helix, read both as a reflection and construction of biologic essentialism, shows the intimate liaison between the place of women in science and the science of woman's place. Whereas Franklin is depicted as obstinate, standing in the way of science's progress, the other women know their proper place in the pursuit of knowledge. They accommodate scientific work by offering their promotional, administrative and seductive qualities. This articulation of suitable tasks for men and women is perfectly in line with the ideology underlying genetic thinking, which often fosters a belief in the 'natural' qualities of men and women. Accepting the claim that *The Double Helix* is a realistic report of how science is done leads to a focus on the position of women in science. Looking at this book in terms of theatre highlights the significance of 'gender scripts' or generally allotted gender roles in science.

Through an ingenious use of innovative plots and a strategic recasting of characters, Watson manages to reframe the image of scientists in general and that of genetics in particular. In his foreword to *The Path to the Double Helix*, Francis Crick states that Watson was not really interested in writing history as such, but that his principal aim was 'to show that scientists were human, a fact only too well known to scientists themselves but apparently not, at that time, to the general public'.[18] Watson frequently praises the amenities of laboratory life, always finds time for a game of tennis or for socializing with his peers. Obviously, his story had to appeal to young people, to raise the interest of bright students in doing lab work, which was often regarded as dull and uneventful.

The Double Helix is also an explicit attempt to liberate this new science from its dubious past, as Watson negotiates a new historical context for the 'new biology'. Except for Erwin Schrödinger's *What is Life*, which Watson mentions as a 'source of inspiration', there is hardly any reference to anything or anyone in the field of biology preceding the discovery of the double helix. Apart from a few grand old men, such as Sir Lawrence Bragg and Linus Pauling, who play a minor role in Watson's drama, DNA research seems to come out of the blue. Watson's and Crick's model for the structure of DNA is not presented as part of a chain of discoveries, starting with Johann Miescher's discovery of nucleic acid in 1874 and refined by Oswald Avery's discovery, in 1944, that DNA and not proteins are the source of genetic material. Instead, *The Double Helix* makes the new genetics appear as a brand new discipline resulting from a merger between biology, crystallography, physics and some other disciplines. Perhaps inadvertently, *The Double Helix* is presented as an origin story: the beginning of a promising new era in a new scientific field.

Rather than contrasting genetics with the compromised field of nuclear physics, Watson associates it with aeronautics and space travel, wiping out all undesirable traces of eugenics. Molecular biology, in 1968, had to compete for money and attention with other science projects, such as the Apollo Space Program. Inherent in this mega-project is its strong appeal to the popular imagination. The Space Program obviously aimed at finding new living space to remedy the immanent 'population explosion', and genetics was now consequently aligned with the Program's unstated goal to ameliorate the 'quality of life'. Implicitly, the author of *The Double Helix* tries to establish parallels with the space adventure: the rhetoric of discovery, exploration and finding the 'secret of life' abounds. Like the race to the moon, the genetic race involves competition and rivalry between teams. Negative images of geneticists tinkering with human life are thus replaced by positive images of astronauts embarking on a trip to discover new worlds: the worlds of inner space, the tiniest components of the body.

The metaphorical alliance of these two explorative expeditions – the race to discover the moon and to understand the structure of DNA – would never have worked had Watson and Crick not promoted their helical model as the 'discovery' that rendered the gene palpable. Like the moon, the gene acquired the status of a physical object that is known to exist, but is as of yet unexamined, waiting to be explored by humans. The moon and the gene, in their opposite dimensions, both signify objects that allow the penetration of scientific instruments, the eyes that make their surfaces visible and their supposed underlying structures knowable. The figurative equation between a journey to the moon and a journey to the inner sanctity of the human body is nicely visualized in the 1966 classic *Fantastic Voyage*.[19] In this movie, scientists magnetically reduce an ordinary sized spaceship containing four surgeons and technicians to such an infinitessimal size that it can be injected in the bloodstream of a famous scientist, who urgently needs to have a tumour removed from his brain. Through the camera, the miniaturized scientists are followed in their heroic endeavours to save the man from the threat of a congenital disease. They remove the malicious cluster of cells – four tiny creatures battling a giant hero called cancer – before they finally escape from the vagaries of biology back to the safe environment of the laboratory. Watson's invocation of moon journeys obviously resonates the popular imagination at that time.

Publication of *The Double Helix* marked an important turning-point for genetics: it retrospectively defined the discovery of the helical structure as a landmark event. Watson carefully shaped the meaning of genetics by presenting his text as the equivalent of historical truth. In that respect, *The Double Helix* shares essential rhetorical features with the original 900-word article hypothesizing the structure of DNA, published in a 1953 issue of *Nature*.[20] The model's structure forms the figurative equivalent of Watson's and Crick's article: both the double helical model and the article propose how four bases in DNA, ordered in two chains wrapped around each other, together form a DNA-molecule. As evidenced by two-dimensional X-ray photographs, the model is presented as an abstract representation of a physical object. After 1953, the model became less a hypothesis of reality than a mirror of reality, as both the model and the article were increasingly referred to as the 'discovery' of DNA. Along similar lines, Watson's autobiography presents a hypothesis of a chain of events that finally led to their scientific claim, but later on his claim came to be looked upon not as a subjective interpretation of events, but as a historical 'fact'.[21] Just as the scientific article in *Nature* is assumed to prove the existence and structure of DNA, *The Double Helix* pretends to record that this is how DNA was discovered. To accept this premise – albeit to criticize its contention – is to ratify its status as a record of facts, rather than a rhetorical construct.

Later popularizations of genetics confirm the status of *The Double Helix* as a 'source text'. In 1971, in an article in *Time*, Watson's book is mentioned as the equivalent of Darwin's *On the Origin of Species* because it 'marked the maturation of a bold new science: molecular biology'.[22] A BBC documentary produced to commemorate the twentieth anniversary of the double helix uses Watson's text as a script.[23] Even the most recent popular re-creations accept Watson's account as an unmediated inscription of reality, and uncritically adopt his plotting.[24] *The Double Helix* has indeed become an 'origin story', a book of record, an assumed historical account of how Nobel Prize winning work was done. In popular media representations, Watson's claim to have discovered 'the secret of life' has taken on almost metaphysical dimensions. In writing *The Double Helix*, he not simply uplifted the status of genetics, but also strategically redressed allegations of geneticists being 'immoral' tinkerers with human life. Watson effectively countered dominant framing of genetics in mainstream media as a suspect discipline, by proposing positive configurations of the gene and the geneticist, and particularly through the use of innovative plots like the competitive race.

SCIENTISTS PLAYING GOD

At the end of *Life Story*, a 1987 BBC docudrama based on Watson's auto-biography, the Crick character, staring at the helical model, exclaims: 'All we wanted was the body and we got the soul'.[25] Body and soul are not the only allusions to religion. *The Double Helix*, as well as numerous other popular accounts on genetics in the 1960s, are replete with metaphysical references.[26] The abundant use of religious imagery and symbols shows how genetics was constructed as a spiritual and moral issue. In popular media stories and a number of non-fiction books, geneticists were scripted in opposition to clergy and religious leaders, who were supposed to evaluate technology's moral defensibility.[27] Whereas scientists were given the opportunity to hail the wonders of genetics, religious spokespersons were invoked to assess the permissibility of scientists 'playing God'. Each of these professional groups is ascribed a specific authority. Scientists have absolute authority over the interpetation of scientific knowledge, while their 'moral contenders' are authorized to judge the potential *implications* of scientific practices. Yet how well does this preferred script measure the animosity between geneticists and clergy against the meaning of genetics?

The title and structure of an article published in *Time* magazine in 1971 illustrates this typical journalistic frame: 'Man into Superman. The Promise and Peril of the New Genetics' devotes half its space to reviewing the latest developments in genetics research, and the other half to the moral dilemmas released by molecular biology.[28] The opening lines of the article are illustrative of the strong moral overtones imposed by the reporter:

> The quantum leap in man's abilities to reshape himself evokes a sense of uneasiness, a memory of Eden. Eat of the forbidden fruit, God warns, and you shall surely die. Eat, promises the serpent, and you shall be like God (50).

The entire article revolves around the question of whether scientists should be trusted with their immense powers, or whether the newest developments in genetics will usher the downfall of man. By rearranging enough molecules to create life itself, the *Time* reporter warns, man will inevitably invoke comparison to the legendary Faust, who attained power over life and death after he had bartered his soul to the devil. 'If the new knowledge is used recklessly, Faustian man of the future may wonder if he, too, has not made a pact with dark forces' (33). But despite these ominous warnings, the article never calls into question the rightness of each position, or the qualification of each 'side' of the spectre. Geneticists, according to *Time*, take moral objections from religious leaders very seriously, and clergy are quoted as saying that they respect geneticists' work, notwithstanding the Christian doctrine that man, in no circumstances, may tinker with sacred values of life and family. Stories in magazines, journals and non-fiction books are frequently larded with quotes from enlightened Catholics and Protestants expressing their confidence in the potential of science to enhance these sacred values.[29]

As long as molecular biologists limit their work to assisting rather than altering human nature, most moralists are willing to adopt a wait-and-see attitude. This comforting conclusion is the tenet of an article in *Newsweek*, summarizing the various religious points of view on the issue of genetics.[30] The reporter quotes a few religious leaders, who agree that geneticists who respect the integrity of living human beings can do little harm. Professional moralists will never allow scientists to drift off into new perplexities, the reporter states, even though he admits that 'Christians have a habit of arriving late on the scene'. Few theologians question the ethics of improving the quality of biological life. The increasing demand for legalizing abortion worries Christian representatives far more than the 'well-meaning intentions' of geneticists.

Despite the opposition between scientists and clergy inscribed in media representations, there seems little evidence to support a 'fierce dispute' over the morality of genetics. In fact, popular media frequently quote theologians or clergy who are in favour of the new genetics, and who actually support the work of geneticists. Quite a few religious leaders, in the late 1960s and early 1970s, attached their theological seal of approval to genetics in the popular non-fiction they published on this issue. In his book *Utopian Motherhood*, former Catholic priest-turned-embryologist Robert T. Francoeur contends that geneticists are God's co-creators in the evolution of new life forms.[31] Humans have been endowed with the gift of intellect, so they are bound to attempt to emulate God's creation. Even though religion has never operated on the premise that man is his own maker, human beings have the right to stipulate what is human, and have always exercised that right. Or, as Francoeur states in his introduction: 'The Creator has somehow shared with us his omnipotence. Having created us in his own image, he now asks us to share with him in the ongoing creation of mankind and man' (viii). His faith in the knowledge and wisdom of geneticists leads Francoeur to believe that the human race can and should be optimized through genetic engineering.

Another moral expert frequently quoted in popular media stories was Paul Ramsey, a Professor of Religion at Princeton and one of the leading Protestant ethicists in the late 1960s. His book *Fabricated Man*, which reveals an unabashed faith in the potential of the new genetics, gained widespread recognition in Methodist circles and beyond.[32] Examining the ramifications of genetic control, Ramsey regards the advances of science and technology as a mixed blessing. A fervent opponent of abortion of foetuses with birth defects, he recommends that society ought to develop a sense of 'genetic duty'. Ramsey agrees with early geneticists like Hermann J. Muller that the human race is heading toward a genetic cul-de-sac. This scientifically based apocalyptic vision serves to prove his point that humans need to be upgraded in order to prevent final 'genetic decomposition'. Hence, genetic engineering aimed at repairing or preventing defects before birth – or, preferably, before gestation – can be considered a profoundly moral activity. Geneticists, Ramsey argues, are the 'Abrahams of science' who assist us in 'overcoming our destinies' (29). With the help of the new genetics, Ramsey predicts that further deterioration of the human gene pool can be prohibited. He supports identification of carriers of congenital defects in order to inform these people about their 'genetic imprudence' if engaging in an act of reproduction, and proposes the introduction of premarital blood-tests to screen for heriditary diseases:

If we require premarital blood tests to protect the innocent or the unknown from being infected with venereal disease, it would be entirely proper as a matter of public policy to require certain premarital genetic tests to protect the innocent and the unborn, and the unknowing partners

contemplating marriage from complicity in a tragic birth they may not
and should not want (97).

Ramsey advocates any practice in the field of genetics that will benefit our
offspring, as long as it does not harm any human beings already conceived. Genetic engineering was thus viewed as the 'science of life' that
might lead to the ultimate prevention of abortion. While clergy did not
radically oppose genetics, scientific principles and practices were actually
deployed to support their ideology of 'life' – a standpoint that later
became the centre of the 'pro-life' controversy. Contrary to the journalistic
scripts informing the public, not all clergy contested scientists' definition
of the new genetics, but rather borrowed their scientific authority to back
up religious views on conception and procreation.

By the same token, however, geneticists and molecular biologists borrowed the discourse of religion to popularize their new scientific ideals
and ideas, as they drew extensively on metaphysical rhetoric and symbols
to present strategically their new paradigms in terms of a 'belief'. In 1958,
Francis Crick summarized a series of rules on the process of DNA passing
on genetic information, and the way it orders up fabrications of new cellular protein, as 'the Central Dogma'. The crucial point of the Central
Dogma is its insistence on unidirectional causality: information flows from
DNA to RNA to protein. The principle repudiates the possibility of external, intra or intercellular environmental influence on genes. The choice of
the term 'dogma', as David Locke has convincingly argued, was deliberate
and not without irony.[33] The very idea of a set of divine laws enabled the
concept of scientists reading God's will writ in nature, like priests reading
his will in the Bible. Biologists thus developed their own set of beliefs,
their own 'book of life' and their own universal laws that could only be
explained by the high priests of molecular biology. Locke goes as far as
comparing the promotion of genetics to the promotion of a secular belief
with religious overtones:

> Still, as it swept on to one scientific triumph after another, molecular
> biology acquired something of the air of a crusade. United in their unre-
> lenting pursuit of mechanistic truth, its members marched ahead waving
> their banner, with its emblem of the double helix, proclaiming their
> motto of the central dogma and shouting that they had discovered the
> secret of life (50).

In the wake of Crick's formulation of the Central Dogma, we can witness
the development of a chain of related religious images. When Jacob and
Monod published their revised model of the DNA structure in 1963, *Time*

magazine promptly reported this theory under the headline 'Three Men and a Messenger'.[34] According to this theory, strands of RNA are 'messengers' who deliver 'messages' from DNA to protein. Religious imagery meshes peculiarly with the preferred communication and language metaphors. 'Messenger' and 'message' pertain to evangelical propaganda as well as to communication. In terms of communication, the 'messenger theory' means the transmission of information from a knowledgeable source to a receptible one. In terms of religion, it conjures up images of the conversion of heathens, initiating them into the secrets of the holy script. Like Christianity, the Central Dogma has its prophets, its messengers and its message – every ingredient for a secular belief.

So the 'opposition' between scientists and moralists that journalists frame, in fact, appears hardly an opposition at all. The professional views of two groups are obviously moulded to fit the binary frame 'for' and 'against' genetics, yet both scientists and religious leaders reassure the audience that each group, provided they acknowledge their mutual respective professional responsibilities, will keep a lid on possible excesses of genetic engineering. The partition of professional responsibilities between scientists and soulkeepers implies that the first group pursues scientific advancement, and the second is appointed moral guardian of society. Nevertheless, both groups actively shape each other's discourse. While Catholic and Protestant theologians attune their religious standpoints to the new genetic principles, geneticists borrow the language and imagery of religion to present themselves as 'helpers of God'. In popular books and magazines, the occupational interests of these groups are thus smoothly aligned, and mutually reinforce their authority by insisting on the distinctiveness of their professional terrains.

AWE AND SUSPICION

In the 1950s and 1960s, journalists working for mainstream media generally endorsed the seemingly natural science–society split by imposing rigorous binary frameworks; to judge from magazine stories, they obviously did not consider it their task to question the knowledge professed by scientists, or the authority of clergy in evaluating the implications of the new biology. Journalists or the media seemed to be viewed, and seemed to view themselves, as 'messengers' for transferring scientific claims and opposite standpoints to a general audience. But if we look more closely at the way in which journalists framed genetics, we may notice that they did not at all fit the (self-)ascribed role of mere transmitters of scientific knowledge. Although they exhibited great respect for scientific expertise, journalists were also steeped in a general public distrust of a new science that might run out of control.

Scientists considered the media's and the public's ignorance in matters of science as a considerable obstacle to a widespread distribution of genetic knowledge. Indeed, journalists never seriously got to the bottom of molecular biology's innovative paradigm. As a professional group, they hardly seemed capable or willing to dispute scientific experts' claims. Part of the problem was undoubtedly a general lack of knowledgeable reporters who were trained in science. Journalists were overwhelmingly educated as generalists, and specialized reporting was still rare in the 1950s and 1960s. Science was hardly covered as news, and there were far fewer outlets for reporting scientific developments than there are today. If covered at all, science was usually approached from a human interest angle. Abstract, theoretical fields like genetics did not lend themselves particularly well to this type of coverage. The sparse reports on genetics that appeared in the late 1950s and early 1960s were hence short on technical details.

A lack of in-depth coverage cannot be blamed entirely on ignorance of the press in matters of science, but may be partially ascribed to scientists' indifference towards public recognition of their work. Reluctant to discuss their projects with non-scientists, they were especially tight-lipped towards the press. The gap between scientists and non-scientists was not simply a perceived one; the ivory tower mentality could be sustained for a long time because government funding for science projects was taken for granted. As Edward Yoxen explains, expectations of support among scientists were such that they 'felt no pressing need to dramatise their work for a popular audience'.[35] The norms of professional conduct strongly inhibited scientists from publicity seeking, as they formed an autonomous, self-regulating and generously funded professional group. Scientific institutions and research laboratories did not voluntarily open their doors to the press, as the press did not put any significant pressure on scientists to 'open up' to the public.

So on the one hand, journalists respected and accepted scientists' authority in matters of genetics, while on the other they gave voice to a general feeling of anxiety and distrust about the consequences of genetic engineering. Media coverage of genetics in these years was characterized by what Dorothy Nelkin has called an 'awe-and-mistrust' frame of reporting: the unconditional respect for scientists often blends in with a disproportionate fear for future moral and social ramifications of science and technology.[36] This Janus-faced framework characterizes most journalistic accounts of the 'genetics revolution' in the early 1960s. A journalist for

Newsweek, for instance, warns of the 'brave new world' potential of genetic engineering, and at the same time reassures us that geneticists are basically benevolent people, intent on finding a cure for the many inherited diseases that plague humanity.[37] Although scientists are 'in control' of their technology, they unfortunately lack control over the implementation of their inventions. A stance typical for the coverage of genetics at this time, scientists cannot be expected to stop working on DNA just because their knowledge may one day be used incorrectly. Hopeful prospects of banning congenital defects are not enough to prevent potential abuse of technology. According to *Newsweek*, scientists have no bearing on future applications of their work, and thus carry no social responsibility with regard to the implementation of 'their' discoveries. Journalists seem to respect and endorse a strict division between science and society, ascribing geneticists a distinct responsibility in matters of science, while locating the power over the use of technology outside science.

The purported professional challenge to journalists is to reconcile the knowledge generated by scientists, whose authority and claims to knowledge they do not dispute, and feelings of unease shared by the public at large. But given the fact that scientists were not easily accessible, and expert knowledge was hard to understand for a layperson, how did reporters ground their evaluations of genetics? The hybrid discourse of promise and concern is not as much fuelled by rational arguments as by emotional appeals induced by images and imaginations. Extrapolation and imaginary projections of future developments on the basis of current experiments were frequently alternated with pure fictional speculations on the potential consequences of genetic engineering. A number of journalistic stories and popular non-fiction books thrived on the insertion of extrapolation and imagination, whether favourable or critical of the new genetics. Two examples may illustrate this typical kind of journalistic framing in the 1960s. D.S. Halacy's *Cyborg. Evolution of the Superman* (1965) resonates a boundless faith in the new genetics, whereas Gordon Rattray Taylor's *The Biological Time Bomb* (1968) warns against its desultory consequences.

Halacy evaluates the goals of the new genetics in the perspective of several analogous developments in contemporary biomedicine, such as body prosthetics, internal organ replacements, chemical drug experiments and in vitro fertilization.[38] Of all new technologies, he considers genetic engineering the most promising one, because it provides a lasting solution to physical defects: 'Rather than patch him up after he is born, why not produce superior men to begin with?' (162). Halacy hails the coming of a genetically engineered race, and the creation of a 'Superman'. He enthusiastically endorses the ideas of geneticist Hermann J. Muller, who propagated in the 1930s the establishment of sperm banks.[39] The author of *Cyborg. Evolution of the Superman* considers genetic engineering to be the only serious possibility to ban congenital disease and perfect the human race. Halacy skilfully disguises projections and fantasies as probable realities; his extrapolations of future applications are consistently supported by quotes from scientific articles and statements from well-known authorities in the field. Exhibiting an unabated faith in geneticists, he trusts them with the awesome instruments to unravel the structure of man's genetic make-up:

> The potential inherent in the secret of life is tremendous. If man can create life, he can also change life ... The tragedies of malformed babies and other evidences of nature running wild may be avoided. Perhaps through DNA we can achieve true happiness for everyone. Our food is a life form and DNA may help us produce more and better food for the growing population of the world and also the astronauts who will colonize other planets. (183)

Instead of associating genetics with suspect branches of science such as nuclear physics, Halacy, like Watson, relates the exploration of genes and cells to scientific projects enabling humankind to fly to the moon and conquer new planets. In doing so, he refers to the immanent threat for overpopulation and the uncontrollability of the 'quality' of human evolution. Not once in his book does Halacy attempt to explain what the new genetics actually entails. Rather than providing intelligible explanation, he rallies support for scientists who work on the cutting edge of science.

Diametrically opposite in its intentions is Gordon Rattray Taylor's *The Biological Time Bomb*. When the book was published in 1968, it became popular overnight; distributed as 'book of the month', it was reprinted many times. Taylor, as the publisher informs his readers on the cover, is an 'able scientific journalist' who has profound knowledge of the field, and can thus be trusted in his judgements. Like Halacy, Taylor evaluates the entire palette of the latest developments in new reproductive medicine, including test-tube babies and cloning.[40] His message is profoundly apocalyptic: tinkering with human life, besides being morally indefensible, will lead to the ultimate destruction of mankind. Unlike Halacy, who blames anxiety about genetics on public ignorance, Taylor whips up these underlying fears by focusing solely on the potential harmful effects of genetic technology.

Taylor's discourse of concern, just as Halacy's discourse of promise, is made up entirely of evocative images and fantastic imbrications. Since the explosion of the atomic bomb, the public's esteem for scientists has been severely damaged, and, as Taylor argues, the genetic revolution will turn out to be as explosive as the nuclear one. Feeding Taylor's doom scenario are stereotypical images of the evil scientist and the scientist unable to control his invention:

> The explosion of the first atom-bomb drove a jagged crack through the superman image. From behind the mask of the beneficent father figure the mad engineer suddenly looked out, grinning like a maniac. As the impact of the biological time-bomb begins to be felt, the haunted look of Dr Frankenstein may gradually appear on the faces of biologists (220).

Rather than supporting his claims with scientific details, Taylor invokes the Frankenstein image to state that genetic engineering is bound to accelerate beyond human control, and that scientists cannot be trusted with the immense powers of technology. Taylor warns his readers that new techniques should be put in a 'biological ice-box' until society is ready for them.

The discourse of concern is very similar to the discourse of promise in its assumptions and narrative plotting. Taylor also endorses a split between science and society; he equally relies on fixed images and imaginations and abundantly backs up his statements with quotes from eminent scientists. He wields the instrument of name dropping, yet his references are never explicitly attributed to a source, and his book conspicuously lacks a list of works cited. On the cover of the book, a publisher's note affirms the integrity of the author, stating that Taylor 'by no means lets his imagination range with the freedom of science fiction' since he 'rigorously confines himself to developments already implicit in research carried out at this moment'. The entire argument, however, is built on the projection of imaginary tales and speculation of future scenarios.

While the authority of scientists over the interpretation of genetic knowledge is fully acknowledged, scientists seem to have little power over their public image. Popular books like *Cyborg* and *The Biological Time Bomb* are remarkably similar in their deployment of narrative strategies. Neither author exerts himself to explain the scientific content of genetics to a general audience; their prime goal is respectively to push or bash genetics by invoking emotions. Quotes from scientists are bent one way or another, depending on the author's stance. And the persona of the geneticist is typically represented through projection of one-dimensional, stereotypical characters: he is either totally trustworthy or thoroughly suspect. A profound fear for 'immoral science', as reflected in attenuated *Frankenstein* and *Brave New World* interpretations, effaces all basic trust in science and scientists. Journalists commonly discuss genetics in terms of its consequences, which are removed from the scientific periphery and instead taken into the realm of morality. Taken 'outside' the boundaries of science, geneticists seem to have as little control over the production of their popular images as they have control over the implementation of technology.

CLONING FOR ETERNITY

One particular aspect of genetic engineering that frequently emerged in the hybrid context of fact and fiction is the issue of cloning. In the scientific understanding of the term, cloning or mononuclear reproduction is not even necessarily related to genetics, but in the popular mind it was strongly associated with bioengineering, stirring up agitated fears and fantasies.[41] In postwar decades, cloning became the focal fixation for the anxiety over the new science and epitome for the way that the whole question of genetic germ material dissolved into spiritual and metaphysical speculation. To religious people, it symbolized asexual reproduction that makes the holy bond of marriage obsolete. Outside religious circles, the potential of parthenogenesis induced philosophical reflections on the nature of human life, and raised questions about the uniqueness and integrity of individuals, the essence of humanness, and the morality of stretching one life into eternity. Against a backdrop of still lingering concerns about Nazi practices and Cold War tensions, the idea of cloning elicits strong fears about its potential abuse by dictatorial regimes who want to extend their power beyond 'natural' life. Cloning is a powerful idea because, as Donna Haraway contends, it is simultaneously 'a literal, a natural, and a cultural technology, a science fiction staple and a mythic figure for the repetition of the same, for a stable identity, and for a safe route through time seemingly outside human reach'.[42]

Mononuclear reproduction is not just a literary anxiety; the 'fantasy' itself harbours the slippery slope between fact and fiction, projection and speculation. Although journalists firmly positioned themselves within the boundaries of factual knowledge, they invariably relied on fictional techniques and resources when writing about cloning. The prospect of genetic technologies apparently appealed to the imagination, and imagination was often the only recourse for journalists who lacked the necessary scientific sources. In the 1960s and 1970s, cloning was still a far-fetched technology, yet journalists discussed the possibility of mononuclear reproduction as if its realization were imminent. Long before the first successful in vitro conception, the idea of extracorporeal reproduction resonated widely through popular culture.[43] The media's fascination with cloning is reflected in popular hybrid accounts that wavered between science journalism and science fiction. David Rorvik, a special reporter for *Time, Science Digest* and *Esquire*, uses the typical awe-and-mistrust approach to envision the prospects of cloning.[44] Looking at the bright side, man can endlessly reproduce himself and thus extend life beyond death. The human race may be optimized through cloning the best and the brightest; subsequently, clonal offspring may be engineered to make humankind adjustable to new environments, such as the moon. The dark side of cloning extends the dangers of eugenics into the new genetics. In order to prevent future carbon copies of Hitler or Stalin, Rorvik suggests establishing a Commission for Genetic Control as soon as possible. Never engaging in outright moralism – prohibited by the conventions of journalism – Rorvik alternates fact and fiction, history and speculation, myths of Superman and the Space Man, to 'prove' the alarming imminence of cloning.

But more than journalism, science fiction formed the apparent playground where the cultural implications of cloning could be discussed, its social permissiveness be probed, and larger philosophical concerns about human integrity be raised. The idea of parthenogenesis triggered reflections on the relationship between technology and society, and the effects of technology on concepts of nature and culture. Science fiction tales on cloning typically take on the form of 'what-if' scenarios: what will happen to the human race if selected individuals can be infinitely reproduced? Cloning may indefinitely extend the life of people in power, and thus endanger the power equilibrium ensured by the intrinsic finiteness of human life. Obviously, a science that is bound to unravel the beginnings of human life holds infinite potential for overcoming the end of life. It is precisely this aspect of cloning that generates compelling questions about a life's 'natural' duration and quality. In the face of endless reproduction of one body, can we still define each of these clones as individuals? How does cloning affect our social structures, if it forces us to regauge the essence of individuality and humanness?

To address these profound anxieties and to reflect on the wider implications of genetic technology, some science fiction writers set their stories in a fantastic space, far removed from empirical reality. They invent an imaginary world where new social patterns have emerged as a result of human cloning. Frank Herbert, in his novel *The Eyes of Heisenberg*, pictures a world in the next millennium which is ruled by 'Optimen': a group of clones who have carefully engineered their own class to become superhumans and rule the world.[45] They are the absolute perfect race – healthy, intelligent, civil, beautiful – yet their superhumanity makes them almost inhuman, as they lack normal emotions or feelings. Their dictatorial control is contested by the Sterries (sterile subhumans and breeders) and the Cyborgs. The difference between Optimen and Cyborgs is that Optimen are genetically shaped, while the Cyborgs have undergone physical alteration of their bioframe, which makes them half human, half machine. The Cyborgs' strategy to obtain power is to steal clones from the Optimen and insert a strain of 'violent' DNA in their offspring, so that the Optimen will be gradually infiltrated by aggressive siblings who are genetically programmed to cause discord and conflict. In addition, the Cyborgs run a 'Parents Underground' which aims at upsetting the biological stability of the inheritance pattern, and at undermining the chemical stability of the molecules in the Optimen's germ plasm. Eventually, this strategy must break the source of the Optimen's power: their ability to live eternal lives.

Religious overtones and metaphysical allusions abound in *The Eyes of Heisenberg*. Its hero Doctor Svengaard, a human geneticist, gradually becomes aware of his own exploitation by the Optimen, whose superior status he helps perpetuate through cloning their leaders. When he is unwittingly involved in the work of the Parents Underground, his eyes are opened to the whimsicality of the Optimen's Central leaders, whom he previously considered Gods: 'To him, Central and the Optimen were the "Primum Mobile" in control of all celestial systems' (101). When he is finally taken hostage into the command centre of the Optimen's territory, he realizes that his world is not ruled by a Creator, but by a bunch of cold technocrats. Svengaard's religious belief in his leaders cools off as soon as he becomes aware of their secular belief in the gene: to Optimen, genes are the essence of a hierarchical social system, as they provide the key to eternal reproduction of life, and thus the key to power. Genetics is unmasked as a religion, but the next step is that religion is overturned by a universal craving for human values. When violence breaks out among the Optimen, triggered by the engineered violence genes, Svengaard follows his professional instincts and human morality, and saves the lives of as many individuals as possible, including those of Optimen. Two Optimen leaders who are saved by Svengaard are so impressed by his actions that they propose to co-operate with the humans. In exchange for human emotion and empathy, Svengaard demands that Optimen give up their claim to eternal life: it is their immortality that wipes out their sensitivity and turns them into vain, soulless creatures.

The spatial layout in Herbert's novel signifies a strictly divided social order. The Optimen inhabit their own continent, an immense city built on top of a vast underground laboratory, in which all the cloning and growing takes place. Sterries and breeders are never allowed above ground; they are banned to the subterranean workforce. This bipartite world is split into upper and lower halves, a division based on laws of oppression. The Optimen control both worlds; the Parents Underground – which ironically operates above ground, in the world of the Optimen – cannot hide from the total surveillance of the ruling class. Yet, in accordance with Heisenberg's law, it is impossible for the Optimen to observe everything at once; the underground operating rooms – the locus of genetic engineering – escape the rulers' eyes. Control over the entire system turns out unfeasible, but the Optimen are blinded by their own belief in total surveillance. What might look like progress is in fact degeneration. It takes a hero like Svengaard to open the eyes of the perfect race to their imperfection. Rational regulation of reproduction does not result in an optimized race, a class superior to ordinary humans and Cyborgs, but instead marks the beginning of degradation into subhumanity.[46]

In *The Eyes of Heisenberg*, the vindication of humanity over unscrupulous scientific dogmas entwines questions of science and power. The new genetics is unmasked as a secular dogma, and criticized for its grand claims to unravel the 'essence of human life'. Herbert uses the parallel between religion and science to argue that control over the duration of some lives, and the mortality of others – whether controlled by deity or by scientists – can never affect the essence of humanness and humanity. It is particularly this aspect of cloning that captured the imagination of science fiction writers who contemplate the philosophical implications of extending human life beyond normal duration. What happens if individual lives can be prolonged *ad infinitum*? Will cloned individuals remain the same even if historical circumstances and society change? If so, what is the unique contribution of an individual to society? These kinds of question surface more poignantly in science fiction stories that are set in a familiar realistic setting. Relating technological developments to actual historical and political contexts commonly heightens the perceived urgency and apparent imminence of technical feasibilities such as cloning.

Ira Levin's *The Boys from Brazil* is set in a recognizable time and space: a postwar Europe, which is still recovering from the horrors of the Holocaust.[47] A fictional version of Simon Wiesenthal named Liebermann finds out from a journalist that some ex-Nazis in Brazil are planning to kill ninety-four men throughout Europe and the USA, yet he does not under stand why these men are targeted; the only common denominator seems that all men are sixty-five years of age, and have fourteen-year-old adopted sons. From then on, the story evolves like a detective novel, in which Liebermann gradually puts the pieces of the puzzle together. Halfway through the book, the reader understands that Dr Mengele, who fled to São Paolo after the German capitulation, has managed secretly to clone tissue from Adolf Hitler's body. Out of this body tissue, he produced in a secret laboratory ninety-four clones of the Nazi leader, who, after having been gestated by an equal number of surrogate mothers in Brazil, were then adopted by childless couples all over Europe.[48] The adoptive parents fit Hitler's social profile: a protective mother who adores children and a father who is fifty-one when they receive the boy. In order to simulate Hitler's familial situation, Mengele had ordered his Nazi helpers to kill the boys' fathers at the age of sixty-five.

After Liebermann has managed to unravel the conspiracy and eliminate Mengele, he discusses the future of the ninety-four living clones with his friends from the Jewish Defence League. Should they kill all the boys who have Hitler's exact genetic profile in order to prevent another Holocaust? Liebermann's friends argue they should, but Liebermann himself decides differently: he destroys the list with the boys' names and addresses. When his friends angrily tell him 'it's Jews like him who let it happen last time', he answers: 'Killing children, any children – it's wrong. Jews did not "let it happen." Nazis made it happen. People who would even kill children to get what they wanted' (234). Liebermann argues it requires more than genetic inheritance to revive Nazism – it would take a replication of social circumstances, people to support Hitler and historical momentum. Yet the last page of the novel predicts an ominous future: one of the clone boys, a gifted artist, is working on a sketch of a man on a platform, mesmerizing a massive audience with his speech.

The teleological relation between technology and ideology can be inferred from the novel's plot. Nazism, inspired by eugenic ideals, caused the Holocaust; the new genetics, if in the hands of the Nazis, will result in a similar disaster. Yet technological determinism does not entirely coincide with genetic determinism. Even though genetic endowment is thought to play a crucial role in the duplication of Hitler, nurture appears to have a distinct impact on the growing of a clone. The boys' living environment is carefully imitated to create comparable circumstances in the boys' upbringings. In the final chapter, the hero of Levin's book openly questions the validity of genetic predisposition as the only factor in the replication of a man's life: you can replicate someone's biographical 'facts' but you can never replicate historical circumstances and social context. The moral of Levin's novel is that it takes more than nature, and even more than nurture, to copy a human being exactly. In spite of science's awesome abilities to continue life into eternity, historical circumstances substantially affect the basic conditions for the development of human identity. Conversely, historical figures may be copied into a different time frame, but it takes more than one individual to repeat history.

A similar view on the impossibility of reproducing identical human beings underlies Nancy Freedman's science fiction novel *Joshua, Son of None*.[49] Although the novel's utopian intent is diametrically opposite to Levin's dystopian fantasy, the questions raised by these books are strikingly similar. *Joshua, Son of None* relates the story of a young physician, Thor Bitterbaum, who happens to be present in the emergency room in Parkland Hospital in Dallas, Texas when President John F. Kennedy is rushed in. On an impulse, the medical researcher takes a scalpel and lifts a small amount of cell material from the president's neck wound, placing it in a Pyrex tube and adding liquid nitrogen. 'While the world was mourning the loss of the particular genetic combination which had just been senselessly murdered, he carried in nitrogen solution the possibility of a perfect replication' (16). Bitterbaum wonders if he can circumvent the moratorium on mononuclear cloning experiments, and secretly embark on a project to clone the president. He manages to find a sponsor, G.K. Kellogg, a rich business man whose wife is unable to bear more children after the birth of their only son.[50] After a few months of experimentation, a geneticist succeeds in cloning the presidential cell tissue, and they find a surrogate mother who, under false pretences, carries the precious embryo to term. Right after he's born, Josh is adopted by Kellogg and his wife; Mrs Kellogg is told that Josh is a product of her husband's extramarital affair. The couple adopts five more children who are all raised by a demanding father and a devout, loving mother. Bitterbaum, who remains an 'uncle' to the Kellogg family, takes every precaution to re-create the family circumstances that shaped JFK's character. He and Kellogg arrange to hire a sixteen-year-old 'nephew' who saves Josh's life in a swimming accident to re-enact a traumatic near-drowning experience in JFK's childhood. While on holiday in Italy with his mother, Josh's sister Ellen is killed in an accident.

As he grows up, Josh's striking resemblance to the late president causes more and more eyebrows to raise. When Josh finally discovers he is not JFK's illegitimate son, but his clone, he refuses to consider himself a duplicate of the former president, instead insisting on his individual integrity: 'A copy is never the original. Why do collectors hunt for first editions and value them?' (241). He decides to seek public office, not as the JFK clone, but as 'Joshua, son of none', and is elected by a landslide.

While insisting on his individuality, he uses his newly won position to defend the rights of clones:

> I don't advocate cloning as a substitute for natural reproduction. But it has a place. It should be considered in the future when we lose a great artist, or scientist, or religious leader. And we must continue through genetic research to attain physical and mental improvement in the gene pool (270).

Because of Josh, cloning becomes the 'popular version of reincarnation'. At his inaugural speech, the inevitable happens: a rifle shot coming from a building splits Josh's head in two. Bitterbaum, for the second time in his life, accompanies a dying president to the hospital, where he takes a scalpel to lift a small amount of matter from the head wound...

Through cloning, life becomes a self-replicating mechanism. The physician, Bitterbaum, understands that, while unable to save the life of the president, he 'had killed death'. In Freedman's version of cloning, scientists succeed in creating eternal life by recreating the body – a notion that naturally appeals to the popular image of genetic engineering as a means to rejuvenation. After all, eternal life is worthwhile only when it can be sustained by a healthy and young body. It is not a coincidence that the president's youthful appearance has become part of the Kennedy myth: in our collective memory, he stays forever young. JFK's mythic aura and the mysterious circumstances of his murder have given rise to myriad fictional and journalistic accounts, contemplating what would have happened had he continued to live.[51]

Both *The Boys from Brazil* and *Joshua, Son of None* embroider on the myth of cloning, either perpetuating the life of the most evil man in history or that of the most legendary politician of the twentieth century. Both authors conclude that 'life' is more than an exact duplication of its molecular constitution. In order to achieve an actual replication of a historical person, one must not only rely on genetic inheritance but also imitate that person's upbringing. Nature and nurture in these novels seem inextricably intertwined. The idea of individuality and uniqueness takes on cultural importance as the two plots unfold. In Levin's book, Hitler's clones are ultimately not considered their master's copies, and Joshua starts a crusade for the public to accept his own personality, refusing to be a reincarnation of his donor. As propagated in *The Eyes of Heisenberg*, the value of every single human life is defined by its uniqueness and finiteness.

All three fiction writers address philosophical questions concerning the nature and permissibility of cloning techniques, as they appear very critical of both genetic essentialism and technological determinism. The 'essence of life' can never be reduced to molecular units, as 'life' is the result of constant interaction between biological and environmental factors. Whether they applaud or condemn the new genetics, these novelists dispute geneticists' larger metaphysical claim to having found the 'essence of human life'. As critics of culture and science, novelists seem to occupy a clearly demarcated discourse, a 'fictional' space from which criticism is directed at scientists, but is articulated in a format that is safely distanced from the 'factual' discourses of science and journalism. Through fiction, authors may question scientists' claims to truth, yet they can hardly corrode their authority.

Between the late 1950s and the early 1970s, the theatre of genetics can be best characterized as scientists' offensive to alter the suspect public image of genetics. Looking at genetics as a public performance, we saw how scientists engaged in a concerted effort to create a new image for a new field. By revamping the presentation of their public personae, inventing adventurous plots and launching alternative metaphors, the 'new biology' tried to dissociate itself from former eugenics. James Watson's *The Double Helix* succeeded in constructing a new kind of interpretive framework for the general public to measure the achievements of science. In terms of scripting, genetics was overwhelmingly framed as a moral issue in popular media stories. Geneticists were ostensibly opposed by a moralist majority of clergy and theologians. However, what was scripted as a dispute seemed anything but a profound schism between scientists and moralists. Religious leaders spoke rather favourably of the new genetics, and used their scientific claims to reinforce their own religious views on matters of life and procreation. Geneticists, for their part, equally drew on religious discourse to strengthen their own public trust. Although each group appeared to respect the other's autonomy and authority, they surreptitiously pickpocketed the other's persuasive images, jargon and imaginative resources.

In terms of staging, we can infer from popular representations that scientists, journalists and fiction writers each claimed a distinctive professional role in the public definition of genetics. Journalists hardly ever questioned the authority of scientists in matters of genetics, in part because they lacked specialized knowledge and access to scientific expertise. Signalling their wavering position between a basic respect for scientists' authority and the public's vestigial anxiety for genetic theories, journalistic accounts often exhibited a mixture of awe and fear. Speculation and fantasy, as becomes clear from this chapter, was not at all reserved for the domain of fiction writers; journalists and scientists equally deployed imaginary sources to shape popular genetic knowledge. Fact and fiction meld beyond distinction in stories on cloning, the trope that epitomized collective confusion and anxiety caused by the new genetics. Allowing a mixture of projection and speculation, science fiction novels on cloning appeared to be the preferred discursive format for expressing philosophical and ethical doubts. As long as the grander claims of the new genetics were questioned from the platform of (science) fiction, the authority and autonomy of scientists in matters of science seemed relatively harmless.

This mutual demarcation of professional authorities profoundly changed when moral objections to genetic engineering gradually gave way to concerns of a very different nature. In the course of the 1970s, scientists and journalists got involved in a discussion about their respective responsibilities: the responsibility over scientific practice as well as control over its unpredictable consequences. The authority over the interpretation of scientific knowledge became the stake in a dispute over the safety of DNA-research. Along with this contest over professional responsibilities, we can see a shift in the way genetics is publicly discussed, as it turned from a moral into a political issue.

3 Biohazards and Bioethics

MONSTROUS MICROBES

Michael Crichton's *The Andromeda Strain*, published in 1969, in many ways epitomizes the transformation from the representation of genetics as a morality play to the representation of genetics as an environmental concern.[1] Crichton's novel relates the story of a lethal organism brought back to earth by a returning space shot, instantly killing fifty people in the village of Piedmont. Six scientists, who are recruited by the government to handle microbal threats caused by outer-space missions, manage to contain the organism identified as 'the Andromeda strain' in a top-secret laboratory. Despite the most stringent containment conditions, the Andromeda strain escapes from the government facilities, fortunately after it has mutated into a benign organism which can do no more damage to the constitution of human DNA and the environment. On the one hand, *The Andromeda Strain* resounds with familiar associations of genetics with space travel; anxieties of Sputnik and Apollo space missions bringing alien life back to earth typically reflect Cold War tensions, equating the invasion of alien organisms with the penetration of communist ideas.[2] On the other hand, the novel introduces a new anxiety: the fear for engineered bugs escaping from a lab, causing unintended mutations in humans or other organisms, and hence threatening the evolutionary balance on earth. In the politicized mood of the 1970s, genetics got annexed as an environmental issue; this new configuration manifested itself in changed images of genetics, genes and geneticists.

The Andromeda Strain was one of the first novels to draw attention to the potential dangers of microbal life forms escaping from a heavily secluded laboratory. Significantly, Crichton does not arouse fears by zooming in on the dramatic consequences of engineered DNA being released into the environment. Instead, his plot revolves around the deliberations of scientists who try to solve the mystery through hypothesizing and testing scientific premises with the help of advanced equipment and the newest theories in molecular biology. In an unusual move for science fiction, Crichton lists an impressive number of scientific sources at the end of his novel. In this way, he tries to convince the reader that the new kind of plot introduced here – containment of manipulated DNA – is based on extrapolation of actual scientific developments. Over the next decade, the word 'biohazard' became a household term immediately associated with genetics, indicating the fear of an accidental, uncontrollable release of microbes.

During the 1970s, the gene increasingly took centre stage, and its public image mutated along with environmental anxieties. After DNA had been identified as the prime constituent of genetic determination, DNA strings were promoted to the status of protagonists. But, as illustrated in *The Andromeda Strain*, these invisible entities could also be cast as unpredictable, mysterious 'things' whose mechanisms as yet eluded the intellectual comprehension of scientists. Genes were not just benign or neutral 'letters' or 'factories' but might also be recast as tiny agents forming a potential threat to the environment. The reification of the gene makes it subject to scientists' manipulations while at the same time rendering it capable of independent acts. Envisioned as an 'alien' object, DNA represented a profound threat to the physical, mental and social organization of human life. In environmental 'scripts', the (manipulated) gene was cast as the enemy of nature, an inherently evil actor which may or may not be controlled by human agents.

Inextricably bound up with this new concept of the gene was the changing configuration of the geneticist. Whereas the earlier 'awe-and-mistrust' framing of genetics led to evaluations of the geneticist in terms of evil or benevolence, the threat of an escaping string of DNA elicited judgements of scientists' responsibility. A common stereotype in journalistic and other popular writing was that of an irresponsible or sloppy lab worker. Perhaps unwittingly, any lab technician blinded by the prospect of a revolutionary discovery might disregard even basic safety requirements. As early as 1969, MIT biologist S.E. Luria warned of possible environmental dangers emanating from working with engineered organisms. In an article in *The Nation*, he concludes that *Brave New World* has come ever closer; Huxley's nightmarish society may be brought upon us unintentionally, leading scientists to follow their instincts regardless of unpredictable long-term side-effects and ecological consequences.[3] Geneticists, Luria contends, should be called to order by the public; they must 'assume responsibility to tell society, in a forceful and persistent manner, what science is discovering and what the technical consequences are likely to be' (409). The term 'responsibility' appeared to be a key-word in the debate between geneticists and their disputants.

Distrust of the new genetics was not grounded in irrational fear so much as in the idea that its practitioners had never been trained or socialized in responsible thinking. As Horace Freeland Judson explains in *Harper's* magazine, the generation of scientists that created molecular biology was at school when physics was the 'master science', and they 'grew up not so much in the shadow of the bomb, but in the shadow of the decisions to participate in making the bomb'.[4] Scientists who pursue new projects do not have bad intentions *per se*, but they seem to suffer from social and political illiteracy. Judson classifies biologists in three types: the 'sorcerer's apprentice' kind of scientist, who is not aware of the possible evil consequences of his research; the 'Dr Frankenstein type', who believes in science-for-science's sake despite public and moral objections; and finally the 'Prospero type', whose research seems benevolent but who does not realize it may be used to other ends. Scientists, according to Judson, generally do not know when and where to stop; the public cannot ignore 'their sweaty palms and the glint behind the spectacles' and should thus carefully monitor their movements.

Irresponsibility of scientists was either explained in terms of ignorance or arrogance, or as negligence and slovenliness. By virtue of his careless behaviour, the 'wild-eyed professor' might create a 'new killer germ', as *Time* suggests, or a 'monstrous microbe', according to *The Futurist*.[5] An engineered micro-organism, after escaping the lab and multiplying itself beyond control, could find its way into human intestines and cause irreversible damage. The idea of an unintended spill of manufactured DNA-strings outside the laboratory captured the imagination of journalists and readers, because carelessness is a human trait typically associated with the absent-minded, forgetful professor. Sloppiness and recklessness, combined with the narcissistic self-centredness of an obsessed scientist, may readily result in an environmental disaster, as we learn from several media accounts. The association with the careless entrepreneur who dumps his garbage everywhere, favouring short-term gain at the risk of long-term environmental pollution, typically reflects the environmental mood of the 1970s.[6]

The gene as an uncontrollable force and the irresponsible scientist are the main ingredients of Greg Bear's science fiction story *Blood Music*.[7] The novel features the prototype of a sloppy, arrogant biologist. Vergil Ulam is a lab worker at Genetron, a fast growing biotechnology firm in La Jolla, California. Driven by a mixture of scientific curiosity and greed, he completely ignores all safety requirements to pursue his own project: the creation of intelligent cells through a combination of information technology and molecular biology. When Vergil's boss at Genetron finds out about his illegal project, he fires him on the spot and orders him to remove all substances from his test-tubes. To save his precious engineered microbes, Vergil injects them into his own blood. Several weeks after his dismissal from Genetron, he notices significant changes in his body: his eyesight has improved, he is much healthier, and his libido has increased. However, once the genetic sequences in his body are switched on, the cells start to function as autonomous units, duplicating DNA-segments that begin to think for themselves and talk to Vergil. These 'noocytes' ('lymphocytes who can think') develop complex brains and multiply into innumerable 'cities of cells'. Finally, his body starts to disintegrate, transforming into an undefinable substance.

Blood Music embodies the idea that a single, reckless scientist can set off a chain of reactions which ultimately affects the global evolution of interrelated organisms. Vergil's careless arrogance is contrasted with the flat characters of the evil scientist and the benevolent medical doctor. Dr Bernard, working for Genetron, represents the capitalist, selfish scientist who smells tremendous profits if Vergil's invention-run-amok can be properly contained, so that public outcry can be thwarted in time. Edward Milligan, a gynaecologist and a friend of Vergil, recognizes the long-term global impact of this local event, and tries to avert the impending disaster by killing Vergil in the bath-tub. His action comes too late to prevent further spread of the monstrous microbes: the noocytes have already started to multiply at an incredible speed, infecting not just the two scientists, but the entire community through the sewer system. Within a matter of days, the larger part of the USA is invaded by noocytes, who change human bodies into a brown, vegetable-like substance that covers entire cityscapes. Even though the USA is instantly sealed off from the rest of the world, the spill of engineered organisms not only affects the ecological balance, but also destabilizes the political equilibrium: the Russians invade Cuba and Europe erupts in turmoil.

Accidents with engineered DNA, as Bear's novel teaches us, may alter the basic composition of all living matter, and consequently affect our conceptualization of 'life'. Several years after Vergil's disastrous experiment, the world seems to have entered upon a new ice age: all land is covered with a kind of phosphorescent seafoam. While scientists in Britain are still figuring out what happened, the monstrous microbes have drastically changed the face of the earth and have transformed the biological make-up of human beings. Peculiarly, nobody seems to have died as a result of the transformation; on the contrary, all people dead or alive have just substantiated in a different way. As a famous physicist finally figures out, trillions of micro-organisms have effaced the distinction between spiritual and physical essence. The world is no longer a spatial entity in the conventional sense of the word. Notions of time and distance have been annihilated; people can materialize anywhere because the noocytes' consciousness is microscale.[8]

Unlike previously discussed science fiction novels on cloning, *Blood Music* does not revolve around questions of potential (ab)use of DNA-technology, but around potential dangers of the experiment itself. Accidents in the laboratory may unintentionally lead to a total reconceptualization of 'life', identity and body. Bear endows the gene with tremendous powers: it is capable of taking over human evolution and redefining human intelligence. Individual consciousness is defeated by genetic consciousness. Some people who appear unaffected by the physical transformation may be able to retain their integrity of mind and body, but they are cut loose from their genetic ties. The surrender of mind and body to an invading army of microbes becomes the equivalent of the world being destroyed by a hostile army. Bear's science fiction scenario locates the cause for a microbal-environmental disaster somewhere between an unscrupulous, reckless molecular biologist and conscious, imperialist sequences of nucleic acid.

Blood Music reflects the public's fear of manipulated genes which, if not properly contained in the lab, pose a potential threat to the environment. The self-mutating, uncontrollable gene and the self-regulated, uncontrolled geneticist become two sides of the same coin: without stringent safety measures, a single scientist might cause a complete evolutionary upheaval. The gene and the geneticist, in this popular configuration, are cast as enemies of nature – a view evidently based on a strict separation between science and nature. Yet an unmanipulable 'natural environment' is as much an *idée fixe* as a 'manipulative science'. These fixed images of the gene and the geneticist betoken the perceived opposition between genetic determinism and social determinism. In the latter view, society and social groups are responsible for, and have power over, technological innovations. Both views partake of the same oppositional framework of nature versus society. The assumption of the gene as an enemy of nature that lies hidden in these texts sharpens the opposition between nature and society. In the wake of the recombinant DNA-dispute in the 1970s, three special interest groups entered the scene: political activists, ethicists and feminists tried to negotiate scientists' interpretation of genetic knowledge to include wider social and ethical concerns in the definition. But before we take a look at which role these groups played in the DNA-controversy, we first need to examine the changing dynamics between scientists and journalists. With the responsibility of the geneticist thrown open to challenge, the question of professional responsibility over the interpretation and distribution of scientific knowledge moved centre stage.

GENETICISTS VERSUS JOURNALISTS

The absolute control of scientists over the interpretation of genetic research was less and less taken for granted, as it became the very stake in a fierce public dispute over biosafety. Yet in spite of common knowledge, it was neither journalists nor special interest groups but biologists who first sounded the alarm bell in the halls of science. In 1973, the Gordon Conference on Nucleic Acids in New Hampshire gave voice to the uneasiness that molecular biologists had felt for some time when discussing recombinant DNA-research.[9] Microbiologist Robert Pollack, from Cold Spring Harbor, was quoted in *Science* as saying: 'We're in a pre-Hiroshima situation. It would be a real disaster if one of the agents now being handled in research should in fact be a real human cancer agent.'[10] Yet the press and the public were only truly awakened in 1974 to the possible dangers of engineered microbes released into the environment and causing environmental damage, when a group of eminent biologists headed by Stanford's Paul Berg published a letter in both *Science* and *Nature*.[11] In this letter, Berg and others proposed a voluntary ban on certain types of DNA-research until the potential dangers of such research had been properly identified, and appropriate safety guidelines had been formulated. The proposal took the form of an appeal to colleagues throughout the world to defer two types of research: those experiments involving the insertion of bacterial genes which confer either resistance to antibiotics or the ability to form bacterial toxins, and those experiments involving the genes of viruses. The danger that bacteria endowed with resistant genes might escape and infect the population still seemed remote, but not entirely hypothetical, warranting a temporary ban.[12]

Only at the onset of World War II, in 1939, had scientists called for a similar moratorium to prevent crucial information about the atom bomb from leaking to the Nazis. As Berg and others proposed in their letter to *Science*, a voluntary ban on DNA-experiments should be acknowledged until the scientific community had formulated appropriate guidelines, approved by the National Academy of Sciences (NAS). That formal discussion took place in February 1975, when over 140 prominent scientists gathered at the Asilomar conference centre in Pacific Grove, California.[13] It took three days and nights of arguments for and against regulation before the organizing committee could come up with a statement listing the safety principles meant to replace the moratorium. The document produced during the Asilomar conference recommended that experiments be ranked in six classes, and that they be conducted in laboratories appropriately safeguarded to their expected degree of hazard. A watered-down version of this document – specifying four classes of risky research – was later published in *Science*; both the NAS and National Institute of Health (NIH) accepted these guidelines and imposed compliance with these norms on all laboratories receiving public funds.[14]

The Berg letter and the subsequent conference rendered responsibility over the safety of DNA-experiments a public issue, but it also addressed scientists' responsibility to inform the public. Due to the sensation the Berg letter had caused in the media, it was clear that Asilomar would provide a unique opportunity to educate the press and the public about genetics.[15] Besides, the conference triggered a symbolic struggle over the demarcation of professional authority and autonomy. As a gathering of scientists, Asilomar was solely devoted to discussions about the potential technical ramifications of biomedical and genetic research; social and ethical concerns never entered the scheme of arguments during the conference. The debate among experts was exclusively focused on judgements of safety, which, they argued, should be based on a solid knowledge of scientific data and extensive lab experience. Since the assessment of environmental dangers required specialized knowledge, scientists secured for themselves the dominant role of decision-making.[16] In the public eye, however, scientists appeared to disagree strongly over the interpretation of scientific data, which were mobilized by both sides to prove either the probability or preclusion of such accidents.

Notwithstanding their disagreements, geneticists and biologists at least exhibited a token of public accountability. They debated – albeit in a limited fashion – their newly acquired ability to manipulate living organisms. By openly discussing possible hazards scientists made a first step to open up their field of expertise to the public. As a result, the awareness emerged that ordinary citizens need to worry about safety in secluded labs. The conference was not only an important gathering for scientists working on genetics, because they openly disagreed over the interpretation of experiments, but also marked a fundamental shift in the relationship between scientists and the general public. Scientists tried to dispose of their public image as an asocial, irresponsible group of professionals, yet, in doing so, they opened up a formerly closed area of knowledge to public scrutiny. Two groups instrumental in this process – the only two groups of non-scientists present at Asilomar – were lawyers and journalists. Lawyers made painstakingly clear that the right of the public to be safe from harmful research supersedes the right of the scientific community to freedom of inquiry.[17] Journalists insisted that the public has a fundamental right to be informed about what goes on in laboratories.

The press played a crucial role in this process of public accountability. Alerted by the overwhelming media response to their letter in *Science*, Paul Berg and the other organizers of the Asilomar meeting took remarkable precautions in their dealing with the press. They seemed to realize that the *perception* of the outcome of the conference would be even more important than the outcome itself. For molecular biologists and geneticists, the stakes at this conference were very high. As Hutton argues in his review of the recombinant DNA-debate: 'If the public could be shown that scientists were acting responsibly in regulating problems emanating from their work, the specter of external control could be restrained.'[18] The conference organizers inplemented a strict press policy. They allowed a restricted number of journalists to the site – sixteen reporters all working for respectable publications.[19] Participating members of the press corps were requested to abide by three rules: not to file stories until the entire conference was over; to attend the conference from beginning to end; and to avoid 'headline stories', instead concentrating on in-depth pieces. In other words, the organizers of a conference on professional responsibility expected to be footed by equally professional journalistic standards. A sensitive issue like biohazard and self-regulation should be approached with appropriate care and sensibility. The conference, in fact, was also an investment in the education of journalists, in the hope that a better knowledge of molecular biology and genetics would lead to intelligent stories in the media.

The press strategy tried out at Asilomar worked as intended. After the conference, major newspapers and journals published detailed and thoughtful reports. Journalists abided by the rules and took their time to discuss complicated risk-analyses with scientists present at the conference. Informed public scrutiny of science appeared to have become a real possiblity. In exchange for being let into science's inner circles, journalists took their time to sort out complex scientific problems, for once relieved from deadline pressure. Asilomar could be regarded as a node of negotiated professional powers – the power over the interpretation and distribution of genetic knowledge. Journalists shared with scientists their access to the general audience, and traded this for access to the interpretation of scientific data.

In the course of the 1970s, journalists had substantially redefined their professional roles; they regarded themselves no longer as stenographers of authorities, but as the eyes and ears of the public. In the decade of political and social countermovements, journalists developed their own sense of professional obligation and responsibility.[20] Many young reporters came to identify themselves with social and political dissent, a conviction which radically changed their concept of professional duty. As watchdogs of society or representatives of the people, they felt it to be part of their vocation to unmask injustice. In light of the Vietnam War, the Watergate scandal, the student movement, the Civil Rights movement and the women's movement, journalists exhibited a growing tendency to act as investigative reporters and muckrakers, and favoured interpretive reporting over so-called objective reporting.[21] Science was one of the last bastions to yield power to public scrutiny by self-asserting journalists, as each started probing the mentally sealed borders between their respective territories.

The changed role of journalists manifested itself most astutely in a movement called 'New Journalism'. A group of journalists called attention to their newly assumed role as representatives of the people; they rejected the possibility of objective reporting and displayed a sense of urgency to question previously unassailable sources of authoritative power. New Journalism is not merely a stylistic innovation, as many have argued, but entails a thorough reassessment of the journalist's professional role. Besides breaking away from the routines and stylistic techniques of conventional journalism, New Journalists openly doubted their position as objective or disinterested observers, and presented themselves as critical mediators between different social groups.[22] A report on Asilomar most typical of these changed journalistic mores is Michael Rogers's story in *Rolling Stone*.[23] Having witnessed the three-day conference in Asilomar, Rogers structures his report as an apparently random sequence of discussions, dialogues and nuggets of information. More than informing the readers about the risks of biohazards, 'The Pandora's Box Conference' proffers a meta-commentary on the interaction between journalists, scientists and lawyers at the conference. Rogers is keenly aware of the changing relationships between these groups, who contest each other's authorities and power. He describes how, in the past, the relationship between scientists and the press was characterized by mutual suspicion and paranoia. Asilomar unveils journalism's newly acquired self-consciousness and assertiveness, making scientists feel uneasy: 'The scientists loved the press when we got Nixon. But when we start hanging around their own backyard, they get very nervous' (39).

Rogers uses a typical feature of New Journalism in freely mixing factual and straightforward information with rhetorical devices commonly found in drama, such as dialogue, vivid description of scenes and an eye for symbolic images. Despite these fictional devices – for instance, a comic description of an E. coli strain escaping from a lab via a tuna sandwich – his characters never become caricatures. A mixture of fictional style and factual report lends itself particularly to a salient depiction of the conference atmosphere and setting, which in turn enhances the symbolic value of a public debate instigated by scientists. Instead of listing the outcome of the debate – the safety guidelines – Rogers stresses the real significance of this gathering: molecular biologists were the first to 'assume voluntarily some measures of social responsibility for their work' (82). In that respect, the reporter concludes, they won a great deal of respect from the public and the press.

The Asilomar conference, partly due to its unique press policy, indicated a symbolic turning-point in the shifting relationships between scientists and journalists. Scientists seemed to recognize that a solid education of the press was a precious investment in better coverage. Journalists, ascribing themselves the role of public prosecutor and judge, came to realize the importance of gaining entrance to the inner circles of science, from which they had been previously excluded. Whereas in the past they had to resort to speculation and fiction, they could now trade insiders' knowledge for fair and balanced reporting. As some critics have argued, this deal-cutting between scientists and journalists led, in the 1980s and 1990s, to excesses in which scientists 'buy' publicity for their research by offering valuable (and often visually attractive) information to the media. From the late 1970s onwards, we can witness an increasing interest on the part of scientists in media strategies. Science journals started publishing instructions on how to translate complex scientific concepts into 'journalese', how to choose apt metaphors and how to offer concrete illustrations to the press.[24] By inviting laypersons into their field of expertise, molecular biologists at Asilomar definitely yielded power to a general public of non-experts. At the same time, they discovered that this token of public accountability was crucial to their reputation and, consequently, to access to public funds.

POLITICIZING GENETICS

Despite the intention of molecular biologists in organizing Asilomar to help the public sort out complex issues and avoid anxiety over improbable or distant developments, the aftermath of the conference showed anything but a reconciliation of ideas between scientists and environmentalists. On the contrary, after the publication of the safety guidelines, the fall-out over biohazard quickly polarized. Environmentalists and political activists appeared the foremost opponents of scientists in the public debate over recombinant DNA; they openly raised questions about possible consequences of engineered microbes released into the environment and the efforts of scientists to contain them. Laboratory workers started worrying about exposure to chemical toxins and the unknown effect of bio-engineered organisms. By 1977, an unlikely alliance between organized religion, environmentalists and labour activists publicly defended the right to laboratory safety for technicians and demanded scientists' political accountability. Doubting the adequacy of the NIH guidelines, several local and national groups took action to block any further recombinant DNA-research in their communities unless a thorough examination of potential biohazards had been completed. Grassroots groups like Science for the People and Friends of the Earth urged local communities to become actively involved in issues formerly thought of in exclusively scientific terms; they wanted scientists to share control over their work with citizens and tax payers. From early 1976 to mid-1977, recombinant DNA-research showed up on the agenda of city council hearings, town meetings, review committees and other political boards around the country, introducing a precedent setting experiment in public participation in science policy.[25]

The most notorious initiative to block DNA-research and impose a genetic moratorium through local politics has become known as the Harvard-Cambridge controversy.[26] In July 1976, the Cambridge city council prevented Harvard and MIT from building a so-called P-3 facility, a laboratory equipped to conduct riskful DNA-research. Banning experiments with a presumed public health hazard within city limits, the city council voted for a three-month moratorium during which a panel of experts and laypersons would examine whether the proposed safety measures were enough to preclude bacterial spills. In January 1977, this Cambridge Experimentation Review Board came up with extra guidelines to supplement the NIH standards.[27] Even though several other local groups followed the Cambridge initiative, not one resulted in the implementation of extra safety measures.

The politicization of genetics also took place on a national level. Senator Edward Kennedy opened a series of hearings on recombinant DNA-research in April 1975, and by the spring of 1977, six Bills were pending before Congress. Scientists who in 1974 had enthusiastically supported the moratorium now faced serious impediments of their work.[28] The American Society for Microbiology and the National Academy of Sciences embarked on a campaign against the Kennedy Bill, supported by the Friends of DNA-group, to spread the anti-legislation tune. In those years between 1976 and 1978 – an episode often labelled 'the recombinant DNA-wars' – polarization between scientists favouring DNA-research and political groups opposed to it intensified.[29] The passionate public debate was constantly fanned by political activists, who succeeded in framing genetics as a political issue. But the pro-DNA lobby ultimately appeared more successful. In September 1977, Kennedy withdrew support from his own Bill. By 1978, the NIH guidelines were substantially relaxed, and responsibility was shifted to local biosafety committees.[30] In the end, scientists had won their political battle, but the debate did not do much to improve genetics' popularity. If anything, the DNA-debate further widened the split between between practitioners of science and the populace around them.

Although the 'DNA-debate' was a controversy over the safety of recombinant DNA-research, it was inadvertently presented in popular media as a rift between scientists and non-scientists. This false impression is both amplified through scientists' self-representation, political activists' self-representation, and the media's use of conventional oppositional frames. Vocal dissident scientists like George Wald and Edwin Chargaff, who argued against building high-risk facilities in Cambridge, were increasingly marginalized by mainstream scientists, powerful players such as James Watson and Joshua Lederberg. Their efforts to alert the public to potential dangers of their work earned dissident scientists the antipathy of many of their peers. As Alan G. Gross recorded in his analysis of scientists' split over the dangers of DNA-experiments, the majority of scientists eventually united behind 'a set of beliefs that emphasized freedom of scientific inquiry without qualification, the practical benefits of research and, by inference, the irrationality of their opponents'.[31] The public conflict among scientists resulted in an ideological backing of Watson's point of view, as dissenting scientists were openly classified as 'incompetents'. Scientists tried to redefine the debate over biohazard as a dispute over scientific qualifications, and one of their foremost rhetorical weapons was to cast their opponents as 'non-experts'. In doing so, they effectively tried to gloss over internal disagreement by disqualifying their dissenting colleagues.

Following the scripts of popular stories, professional roles seemed to be aligned with ideological positions: scientists were invariably cast as 'pro-DNA' whereas non-scientists were invariably positioned against it. The 'recombinant DNA-drama' seemed to allow for two strictly defined roles of protagonist and antagonist. Both 'sides' fed this simplified presentational format through active contributions to popular media. James Watson, heading the pro-DNA-lobby, openly resented scientists who put the weight of their authorities behind the Asilomar guidelines, so that safety restrictions threatened to become enacted into formal laws. In a clever rhetorical move, Watson labelled his opponents 'left-wing activists', and put them on an equal footing with a conservative government they despised, and which had just been dealt a deadly political blow by the Watergate scandal: 'By now, so many top figures have defended the [Asilomar] guidelines that, as with the Vietnam fiasco, none is willing to admit that he has been a naive fool.'[32] Watson's 'defence of DNA' revealed the successful strategy by the DNA lobby to bury Asilomar. He managed to contain the debate to technical matters of safety and laboratory hazards, refusing to address the more profound concerns about genetic engineering that his opponents raised.[33] Armed with technical expertise, defenders of DNA-research conquered the impending threat of legislation by accusing their opponents of ignorance, as they lacked empirical proof to warrant warnings of environmental spills. Watson *cum suis* not only reclaimed their authority over the interpretation of genetic knowledge, but reconfirmed the ostensible split between science and society in the process.

However, the pro-DNA lobby was not alone in acting upon this oppositional scenario. Anti-genetics activists often proudly presented themselves as non-scientists. Just as Watson was regarded the representative of the DNA-lobby, the anti-genetics movement was personified by Jeremy Rifkin. From the mid-1970s on, Rifkin manifested himself as a professional political activist whose purpose in life was to block any research into bioengineering. In order to attain his goal, Rifkin acquired a black belt in all the martial arts of political activism: manipulating the media and the lecture circuit, filing lawsuits, organizing petitions, canvassing politicians and mobilizing grassroots movements. When his group of activists forcibly took over a National Academy of Sciences forum on recombinant DNA, his motivation perfectly seemed to fit the mood of the anti-establishment counterculture.[34] Even after the avowed victory of the pro-DNA-lobby in 1978, and the subsequent relaxation of NIH guidelines, Rifkin only stepped up his political struggle to fight bioengineering experiments on both local and national levels. His Washington-based Foundation on Economic Trends developed into a professional special interest lobby, whose attorneys are on a constant alert to file lawsuits against local and national governments, companies and universities. In the course of two decades, Rifkin's organization faced numerous scientists and university officials in court to dispute safety regulations of DNA-research.[35]

A perennial critic of modern technology, Rifkin has been dubbed everything from a modern-day Luddite and a demagogue to the 'Abominable No Man'.[36] Typifying the contemporary politicized mood, he radically opposed genetics research while emphatically presenting himself as a non-scientist, a layperson who is nevertheless qualified to discuss the ramifications of genetics research. Rifkin asserted that society, rather than the scientific community, should regulate the emerging field of biotechnology. In every possible way, he contested the authority of scientists over the interpretation of scientific claims, thus arguing that the evaluation of science belongs to the public domain. Even though Rifkin's immediate goal in filing all these lawsuits was to postpone experiments and regauge existing safety standards, his long-term goal was to open up scientific practices to public scrutiny, initiate public debate, and, by implication, affect the public's understanding of genetics.

Like many popular media, the journal *Science* covered Rifkin's manoeuvres as an ongoing political battle between the scientific community and its relentless foe. Indeed, Rifkin constantly managed to move centre-stage because he knew how to suit the convenient journalistic framework of ideological adversaries. To the great annoyance of biologists, Rifkin unabatedly succeeded in mobilizing the media to get his points across. According to *Science*, he 'hangs NIH on a technicality of the review process in order to force debate on the larger issue of human genetic engineering and its potential misuse'.[37] Scientists lamented that Rifkin bombarded the media with press releases to ensure good coverage, so he could fight his battle in an environment were he feels most comfortable: in the spotlight of journalism. The popular press, whether liking Rifkin's views or not, could not ignore him. While scoffing at his unyielding fanaticism, they openly admired his rhetorical fluency and media sophistication.[38] In the DNA-controversy, 'the gadfly' or 'media maven' (as Rifkin is mockingly called) properly understood the impact of media on a public debate and perfectly knew how to exploit publicity channels and journalistic frameworks.

Rifkin contested not only the responsibility of scientists for the interpretation of genetic knowledge, but also their control over the distribution of information. He showered the media with his own articles, and magazines and journals eagerly picked up his writings.[39] In the late 1970s, so-called advocacy journalism was on the rise, as evidenced by an explosion of new politically oriented journals. Among professional journalists, there was a noticeable drive to take up arms and assume social responsibility. Science journalists in particular no longer felt obliged to transmit purportedly 'objective' information, and increasingly paid attention to the political and social implications of scientific discoveries and technological innovation.[40] Although his legal manoeuvres drew most attention from the scientific community, Rifkin was as skilled in fighting with arguments as with images. He published a number of popular non-fiction books in which he launched new metaphors and images for genetics.

Algeny – the title of one of his popular books – is what Rifkin calls an appropriate metaphor for bioengineering.[41] Whereas 'alchemy' evokes images of a scientist trying to process metals into gold, 'algeny' refers to 'scientifically changing of an organism by transforming it from one state to another' (17). An algenist views the living world as life *in potentia* just as the alchemist viewed cheap metals as gold *in potentia*. Although its title suggests otherwise, the book largely consists of an extensive critique of Darwin's ideas. Rifkin rejects evolution theory as an unwarranted hypothesis, a formulation of pure conjectures ungrounded in empirical evidence.[42] The promotion of the survival-of-the-fittest ideology, according to Rifkin, served the bourgeoisie and further oppressed the lower classes. Part of its popular appeal is probably that *Algeny* presents an overarching cosmology rather than a jeremiad against bioengineering. Unlike previous anti-genetics crusaders, such as Taylor in *The Biological Time Bomb*, Rifkin does not just speculate on the scary ramifications of genetics' applications. He lards his argument with concrete images to illustrate the 'relentless exploitation of the environment' and anchors his objections to genetics in a broader social critique.

Journalists eagerly picked up this new ideological battle and revamped previous opposition between moralists and scientists to fit the new binary oppositions between scientists and political activists. Mainstream coverage of genetics and molecular biology overwhelmingly assumed the convenient 'for-or-against' framework, as evidenced by a background report in *Time* in 1977.[43] The journalist pays equal attention to potential perils of genetic manipulation as to possible beneficial purposes of engineered DNA. The danger of one sloppy scientist causing a disaster is contrasted with the spectre of local governments imposing crippling restrictions on scientists' freedom of inquiry. The story is balanced in a typical journalistic interpretation of the word: balance consists of a simple opposition of two ideological stances, polarizing scientists and social groups. Despite obvious support of well-known biologists for the environmentalist position, the split between scientists and non-scientists (or anti-scientists) is rigorously enforced. From that same article, however, we can also infer how these assumed opponents strategically redefine the goals of genetics to meet mutual objections. Pro-DNA scientists are quoted as saying that DNA-research may produce micro-organisms that 'gobble up oil spills', thus countering anti-DNA arguments which defined engineered genes as a potential threat to the environment. Rather than a rigorous split, there is a continuous exchange of arguments and images between various groups.

Although the pro-DNA lobby 'won' the debate on biohazards, averting safety restrictions on laboratory research and government regulation, political activists succeeded in another way: they managed to bring social and ethical aspects of genetic engineering to the public's attention. Including ethical considerations in what was known as a scientific paradigm, they redefined the parameters of public discussion. During this process, the lines between professional responsibilities were seriously redrawn. The power of scientists over the interpretation of genetic knowledge was no longer self-evident, but neither was the authority of journalists over the distribution of knowledge. Increasingly, special interest groups learned how to play the media and use all publicity channels available to influence public opinion, at the same time that journalists discovered their new professional vocation to probe the social implications of science and raise questions of ethical responsibility.

THE CALL FOR ETHICS

A second group that seemingly opposed dissemination of genetic paradigms was the emerging group of bioethicists. As ethical concerns among the general public were mounting, the call to include ethics in the study and practice of the new biology became louder. Many observers of science doubted whether biologists and geneticists were sufficiently educated to raise such questions. As Arno Molutsky argued in an article in *Science* (1974), those trained in the sciences have taken 'a more pragmatic, possibly shortsighted, view of these developments'.[44] Academic experts outside biology and medicine, such as sociologists and philosophers, were better trained to assess current scientific developments in view of future applications. Molutsky and others therefore favoured the advent of bioethicists, a new group of professionals who combined expert scientific knowledge with philosophical flair, to become 'the shamans, the priests of the new age of biotechnology'.

As a new special interest group, bioethicists had to carve out a niche for themselves inside the academic hierarchy, and position themselves *vis-à-vis* science. Up to the 1960s, the need for moral testing of genetics had been fulfilled by church leaders and theologians, who were firmly located outside the university walls. A rapid secularization of society seemed to have left a void. Bioethics promised to fill this gap and replace abstract principles and dogmas by empirical reasoning, grounded in secular humanism and motivated by compassion and a sense of justice. Joseph Fletcher, one of the leading bioethicists in the 1970s, resolutely declared the end of moralism, in return for social responsibility and common sense:

> Once God was the final court of appeal in morality ... All of that is changed. Churches will either be transformed along with ethics, or die. Careful thinkers are convinced that religion in the age of science cannot be sustained by the assumption of miraculous events abrogating the order of nature ... The position now is that men, not God, are the ones who are 'abrogating' natural processes.[45]

Unlike religious leaders, whose faith prohibited a balanced consideration of issues like abortion or conception outside the womb, professional ethicists did not dodge these questions. Ethical issues raised by wholesale genetic engineering in fact touched the core of what were once religious taboos. June Goodfield, a science journalist interested in bioethics, inventoried some of the poignant issues in 1977.[46] Should scientists do everything in their power to save the lives of foetuses with congenital defects who would formerly have died a natural death upon birth or shortly thereafter? Should we allow the possibility of women lending their wombs to grow and gestate children who have a different genetic mother? Ethical questions like these, as Goodfield explains, are not simply academic or far-fetched; rather than rejecting these issues on religious grounds, people are urged to use their ethical imaginations, to think through in advance the consequences of certain experiments.

But with whom lies the responsibility for ethical thinking: with scientists, bioethicists or the public at large? From Goodfield's as well as others' popular books on bioethics of the 1970s, we can distil a number of recommendations to stimulate the discussion of ethical concerns. Scientists were urged to take time out to think through ethical questions. However, without proper training in ethical reasoning, biologists and geneticists could not be expected to engage in sophisticated ethical reasoning. According to Goodfield, it was pivotal to implement courses in bioethics in the regular curriculum of scientists, so that ethical reasoning would become part of their professional attitude. Institutions like universities or academic organizations had an obligation to society to raise ethical concerns, to encourage public participation and provoke wide-ranging discussions about ethics. At an even more abstract level, society needed to develop a sense of ethical awareness with regard to reproductive issues – a 'collective ethical sensibility' or 'new moral climate in which social sanctions alone may be sufficient to impede the wholesale inhuman application of a technology'.[47] In addition, bioethicists recommended having an international committee of experts explore the ethical and social issues raised by the new biology.[48]

Bioethicists made a concerted effort to introduce ethical thinking into science, and to incorporate ethicists into the academic ranks alongside scientists. Gradually, bioethics found its institutionalized niche in the form of official committees and academic programmes.[49] Concomitantly, a number of scientists applauded and stimulated the incorporation of ethics in the scientific curriculum. Yet despite these mutual overtures, mainstream media classified bioethics as a counterweight to genetics. The voice of ethicists was commonly pitted against unscrupulous scientists, who did not have the slightest interest in questions of a social or ethical nature, and who only regarded these concerns as an uncalled for slowdown of their work.[50] Indeed, many scientists complained that the press kept nagging them about social and ethical responsibilities, but at the same time they were also concerned about their reputation. After the unpalatable episode of the DNA-wars, geneticists' image had become tainted with misgivings and conjectures. Although recognition of ethical issues would undoubtedly ameliorate their public esteem, they feared that more self-examination and ethical reflection would lead to more restrictions on experimental research, just as happened in the aftermath of Asilomar. An image of geneticists as responsible, ethically considerate human beings, though instantly desirable, could probably result only from a long-term public relations and education campaign.

The question of who is responsible for the injection of ethical considerations in the theatre of genetics was not only addressed by various special interest groups, but was also reflected in the dynamics between journalists and scientists. While scientists, reluctantly or eagerly, included ethical concerns in their definition of genetics, journalists increasingly regarded as part of their professional duty bringing ethical concerns to the public's attention. An interesting illustration of these shifting professional roles is provided by David Rorvik, the science journalist whose stories on cloning frequently appeared in popular scientific magazines of the 1960s. In the aftermath of the recombinant DNA wars, Rorvik published a remarkable book: *In his Image. The Cloning of a Man.*[51] It was allegedly a non-fiction account of a cloning experiment witnessed by Rorvik himself. The publication of *In His Image* was surrounded by secrecy. Months before the actual release, the book was announced as the revelation of a revolutionary experiment in genetics that would upset the world. Publisher Lippincott reassured the public that the events described by Mr Rorvik, whose credentials were highly praised, had actually happened. In a press conference, the author announced he was not allowed to reveal the names of the scientists involved in this experiment. In anticipation of the book's publication,

Newsweek seriously discussed the imminence of cloning, quoting scientists from opposing scientific convictions.[52]

In his Image recounts how Rorvik is approached by a multi-millionaire called Max, who asks Rorvik to find a scientist who can help him obtain clonal offspring. Finding a genetic engineer who is willing to embark on such experiment proves to be the easy part, because many scientists in the field are frustrated by guidelines hampering their experimental research. Max equips the genetic engineer – whose pseudonym is Darwin – with an up-to-date laboratory on a remote, exotic island. Rorvik is invited to witness the experiment as it proceeds. In light of the 'true' content of the book, this invitation seems rather puzzling: the millionaire, who wants at any price to remain anonymous and prevent premature publicity, seems still eager to have a journalist present to witness the entire scientific procedure. Only when the cloning experiment finally results in the birth of Max's son is Rorvik allowed to publish his account and inform the public of an important scientific milestone.

Rorvik seizes the opportunity of the invitation to raise both ethical and professional concerns. First of all, he wonders whether genetic engineering and especially cloning are ethically acceptable. In an unusual step for a journalist, he exhibits his doubts by reviewing arguments pro and con, quoting respectable scientists and bioethicists. Rorvik decides that his professional curiosity, in this case, wins out over his ethical inhibitions, as he goes along with Max's conditions. The very questioning of his position with regard to genetics betokens Rorvik's new conception of professional duty. In the first two chapters of his book, the author presents himself as a responsible journalist, whose role is that of a promoter of public debate:

> It was hard for me to conceive of myself in this new role. The nature of journalism had changed remarkably over the years. The old standards of objectivity, in which the reporter dutifully listed the who, what, why, where, and when of an event and kept his or her impressions and feelings entirely out of the prose, had been found wanting. 'Interpretive' journalism, it was believed by many – and their arguments were often persuasive – was needed to give the 'truth' three dimensions (74).

Insisting on his role as a representative of the public, Rorvik refuses the money offered by his patron, so he can remain independent. His tactics are reminiscent of New Journalism as he comments on the process of reporting by openly guiding his readers through the decisions he has to make as a journalist, thus assuring them of his professional integrity.

When Rorvik's account turned out to be a hoax, several critics cried out against 'the new mores of journalism'.[53] For Rorvik, however, his choice to publish fantasy as a non-fiction account was both a rhetorical and a pub-licitary strategy. *In his Image*, despite the fictionality of its story-line, con-tains a thorough review of the latest developments in the field of genetics. Rorvik constantly authenticates his assessments by referring to profes-sional journals and actual interviews with renowned scientists. Notwithstanding its fictional content, *In his Image* may be less fantastic than non-fiction accounts like Taylor's or Halacy's. By combining factual references with fictional devices, Rorvik enhanced the idea that the events described, if not entirely true, could easily become reality. As in the case of New Journalism, the mixture of factual report and literary style is deployed to raise the public's interest in ethical issues. Rorvik's intention, as stated on the last page of the book, is not so much to inform people about the latest developments in the field, but to elicit discussion:

> I entertain absolutely no expectation that anyone, scientist or layman, will accept this book as proof of the events described herein ... I hope however, that many readers will be persuaded of the possibility, even the probability of what I have described. And if my book increases public interest and participation in decisions related to genetic engineer-ing, then I will be more than rewarded for my efforts (207).

This paragraph reveals the intent of Rorvik's strategy. Publishing the book as non-fiction rather than fiction, he wanted to increase the potential range and impact of his claims. Had he raised ethical issues in the format of fiction, his claims could have easily been dismissed as 'fantasy'.

The need to include ethical concerns in genetic theories and practices was also addressed in the many science fiction books on cloning published in the 1970s. As illustrated in the previous chapter, novels like *Joshua, Son of None* and *The Boys from Brazil* indirectly raise questions of the permissibility and ethicality of extending human lives beyond normal duration. Pamela Sargent's *Cloned Lives* explicitly ties in questions of bioethics with scientists' fear of limitations on experimental research.[54] The story starts in the year 2000, when the moratorium on genetic engi-neering experiments, imposed in 1980, is about to be lifted. Geneticist Hidey Takamura has always hated the moratorium ('which has deprived the world of thousands of talented biologists') and, on the very eve of its expiration date, he clones his friend Paul Swenson, a famous astrophysi-cist. After the birth of five clones has leaked prematurely to the press, a wave of sensational reporting causes public outcry – resulting in a new moratorium. Hidey is strongly resented by his fellow scientists, whose research is again curbed by the government. One of his colleagues reproaches Hidey that if he had waited a little bit, 'people in biological research could have taken the time to educate the public, get them used to the idea of possible experiments' (67). A public relations campaign, his colleague argues, could have pointed out the benefits of genetic research that was carefully controlled. But Hidey objects that, if science had to wait for a consensus every time something new was tried, 'we would still be living in trees and eating raw meat' (67). Resonating with the tenor of media stories of the 1970s, genetics is framed as a discussion over scientists' freedom of inquiry versus the public's demand to participate in the evaluation of science.

Cloned Lives argues the ethicality of cloning but also the ethicality of sensational press coverage. The author strongly rejects dominant beliefs that parthenogenesis results in the duplication of identical human beings. None of the five clones – four sons, one daughter – turns out to be a carbon copy of Paul Swenson, but each has his or her own distinct individuality. Every one of them represents a feature of their father's versatile character: Edward grows up to be a mathematician and gifted violin player, Jim is a poet and writer, Albert becomes an astrophysicist like his dad, Michael an engineer and Kira grows into an eminent doctor and biologist. The problems that these five clones face in the course of their lives seem less induced by the fact that they are clones than by their frequent public exposure. The five children are constantly sought out by the press to exemplify the dangers of genetic engineering, but in fact they are living proof that cloning does not necessarily produce monsters or social deviants.

According to Kira, the only way to diminish public suspicion of geneticists is to fully incorporate ethics into science. Kira in particular epitomizes the breed of new scientists who are not only trained in several scientific disciplines but also in bioethics. A perfect example of the responsible scientist, she counterbalances the stereotype of the sloppy lab worker. Ultimately, Kira and Hidey succeed in lifting the moratorium on genetics by slowly changing the public's perception. The moratorium, she muses in one of the last chapters, has long violated the spirit of free inquiry, but eventually the long process of public education bears fruit. *Cloned Lives* reflects scientists' awareness that a serious consideration of ethics is a crucial tool to a favourable public image, and may avert impediments to scientific research. Kira and Hidey impersonate the view of the pro-DNA lobby: genetics is unduly politicized, resulting in unnecessary guidelines and crippling restrictions. The importance of public support is dawning on molecular biologists who personally experience the consequences of government interference with their work. Incorporating ethics and educating the public appear vital investments in a future public image;

ethical and social responsibility, rather than being a hindrance, will eventually prove to be an indispensable ingredient of a more positive public image.

GENES AND GENDER

In the course of the 1970s, environmentalists, labour activists and left-wing political groups had primarily raised issues of biohazard and safety. Bioethicists replaced clergy in their assessment of moral dilemmas caused by genetic engineering. A third group that entered the staged play of genetics was the women's movement. During the 1960s and 1970s, a number of issues previously considered private affairs had moved to the political arena: abortion, reproductive rights, contraception and new reproductive technologies. The concept of genetic engineering aroused different anxieties for women than it had for men. The realization that 'man-made life' should be taken literally called attention to the gendered nature of reproductive technologies in general, and of genetic engineering in particular. Prospects of advanced forms of genetic manipulation raised anxieties about how they would affect women's reproductive future.

Like political activists, environmentalists and bioethicists, feminists were commonly cast as opponents of genetic engineering, allegedly forming an obstacle to the proper distribution of genetic knowledge. Indeed, quite a few representatives of the women's movement spoke out against genetic technologies which they feared would make female reproductive functions obsolete. In a 1976 article in *Ms* magazine, for instance, several alarming gender-specific ramifications pass in review.[55] Women could be degraded to cattle, used as brooding machines to gestate genetically altered offspring, and gender preselection could lower the status of females and adversely affect the birth of female babies. Familiar *Frankenstein* and *Brave New World* plots were invoked to underscore the potential dangers of genetic engineering. But not all feminists in the 1970s shared this dystopian view. Shulamith Firestone, in her influential book *The Dialectic of Sex*, argued that genetic engineering combined with other new reproductive technologies might lead to the liberation of women.[56] Women should welcome the possibility of splitting biological and social motherhood, Firestone contended; genetic engineering would prohibit them from being biologically enchained to their offspring, and thus offer more opportunities for social equality.

It may be no coincidence that feminist interpretations of genetic engineering and its consequences mostly materialized in the form of science fiction. Ever since the early nineteenth century, the genre has been used by feminists to criticize and reconceptualize contemporary trends in science and technology.[57] Yet although most feminists used the imaginative space of science fiction to either criticize the contentions of genetic determinism or to proclaim the vindication of social determinism, the interrelation between genes and gender was not exactly evaluated univocally. In some novels, environmental concerns were linked to gender-related inquiries, as 'natural' roles of men and women were used to account for (ir)responsible applications of technology. Other feminist novels foregrounded the gendered deployment of technology in a society that is hierarchically structured by sex and race. And finally, some feminist science fiction authors attempted to unravel the subtle interwovenness of scientific instruments and social arrangements. Three novels of the 1970s may illustrate these diverging feminist projections; Kate Wilhelm, Marge Piercy and Naomi Mitchison all reassessed the concepts of science and nature, as they envisioned a future where gene technologies inform reproductive policies.

Kate Wilhelm's *Where Late the Sweet Birds Sang* (1974) is unambiguous in her dystopian conception of a future society dominated by new reproductive technologies.[58] Characters Molly and Ben are the last real humans living in a post-holocaust world, where everything is destroyed by nuclear radiation. To save humankind in the face of environmental disaster, a handful of people attempt to rescue the human species by resorting to cloning. Women undergo conditioning to become breeders, as they are given hormones and stimulants to upgrade their reproductive capacity. Breeders are used as 'hosts' for clones of the best and brightest of men, to ensure a continuing population of capable adults. Women are dehumanized to serve as receptacles for clone-foetuses that are not genetically their own; their bodies are exploited to ensure the genetic future of the country. When Mark – Molly's and Ben's natural son – is born, he is destined to become a social outcast in the small community where all of his brothers and sisters are cloned from tissue. All children look alike, except for Mark, who exhibits typical features of humanness: a love for the arts, emotion, an understanding of nature, and empathy. When Mark sets out to destroy the laboratory, because he is appalled by the practices of regulated reproduction, the community decides to eliminate him. Yet Mark escapes, taking a few pregnant breeders with him. In the end, they manage to establish a new society which valorizes feminine values and natural reproduction.

Wilhelm's configuration of the social implications of genetic engineering concurs with Jeremy Rifkin's and other Luddite forecasts. The destruction of the laboratory and its equipment symbolizes a visceral dislike of technology. Feminist resistance to genetic engineering, however, goes beyond a mere objection to the dangers that it poses to the earth's ecological balance. Manipulation with genes also threatens to upset the 'natural' reproductive process, taking away from women the power to conceive and give birth, and turning them into breeders for a 'superior' species. Wilhelm's novel, more than anything, asserts an intricate relation between nature and femininity. The back-to-nature philosophy informs the superiority of 'natural', unmediated reproduction, which in turn rests on an assumed opposition between nature and technology, and, by implication, between women and men.

An antidote to *Where Late the Sweet Birds Sang* is Marge Piercy's well-known novel *Woman on the Edge of Time*.[59] Piercy's ideas of a reproductive utopia obviously emanate from Shulamith Firestone's *The Dialectic of Sex*, on which she elaborates in a fictional setting. Protagonist Connie Ramos, a poor Hispanic woman from New York, is about to be confined to a mental hospital when she is contacted by Luciente, a plant geneticist from the future. In the year 2137 Luciente takes Connie on a tour to Mattapoisett, a village in Massachusetts, where babies are produced by public consent: every time a person dies, a new baby can be taken out of the brooder – a futuristic room in which embryos are kept in bottles. Not one baby in Mattapoisett has a demonstrable pair of parents: 'We're all a mixed bag of genes' as Luciente explains (100). The purpose of this gene-blending programme is to eliminate racism and sexism. Genes of black and white people are mixed beyond recognition, to break the bond between genes and culture. And since fertilization and gestation take place outside the female body, reproductive capacities are no longer related to sex. The advantage of a genetic engineering programme is that no single person is biologically tied to one single child, so everyone, male or female, can nurture a child. In Piercy's view, nurture is infinitely more important than nature, as genetic composition is merely the random outcome of a scientific game of dice.

Science, in Piercy's utopia, is a political and democratic process in which all citizens participate. Genetics regularly tops the agenda of local councils in Mattapoisett – resonating Cambridge's local debate on biosafety. Like environmental safety, the genetic make-up of the Mattapoisett population has been decided at city council meetings. As Luciente informs Connie, there is an on-going discussion between the Shapers and the Mixers. Mixers only want to eliminate genes linked with disease susceptibility, while Shapers favour breeding for selective traits. When Connie raises doubts whether laypersons can actually make informed decisions about science, Luciente objects that scientists' unwillingness to allow public participation in the past proves that they cannot be trusted with the powers of science. Eventually, it dawns upon Connie that scientists are not the infallible experts that society takes them to be. In order to prevent the inhumane and dangerous brain surgery a team of scientists and doctors are planning to perform on patients in the mental hospital in her real world, Connie drops a lethal poison in their coffee. This act of sabotage can be interpreted as a symbolic start for a citizen's intervention in the course of scientific 'progress'. The motive for Connie's revolt is a sneak preview of a dystopian future, disclosed to her when she accidentally gets lost while time-travelling to Mattapoisett. In this horrifying scenario, everyone is engineered to fit a genetic profile; hierarchy is strictly enforced so that the rich have all the benefits of good food and old age, while the poor are kept locked in separate apartments to be exploited. It is the extreme opposite of Mattapoisett, a warning of what might become reality if contemporary social arrangements are projected onto a future where sophisticated technology determines people's genetic destination.

Both Kate Wilhelm's dystopian scenario and Marge Piercy's utopian projection reflect contemporary 'politicized' discussions on genetics. In Wilhelm's view, genetics is inherently oppressive, and, if in the hands of men, will only lead to a reinforcement of existing inequality between the sexes. Piercy, on the other hand, predicts that if science can be democratically organized, and if genetics is conceived of as a political rather than a purely scientific issue, 'we the people' can harness genetics to erase social inequalities of sex and race. *Woman on the Edge of Time* is not, as some critics have argued, an unbridled panegyric to science and technology. In contrast to Wilhelm, Piercy imagines a profoundly different concept of science: in her view, science is not an activity that is performed behind closed doors, but a process of democratic decision-making. Whereas in Wilhelm's imagination, all power is in the hands of scientists, in Piercy's view the power over scientific knowledge can and should be taken over by the people. Piercy equally enhances the split between nature and nurture, as she favours social determinism, and accepts the absolute hegemony of society and nurture in the shaping of human individuals and society.

Somewhere between Wilhelm's rejection of genetic technology and Piercy's cautious embrace of it, we can position Naomi Mitchison's *Solution Three*.[60] In her science fiction novel, Mitchison questions the unconditional adherence to scientific dogmas underpinning social structures, even if genetic engineering demonstrably advances a just and stable society. She envisions a state where turmoil, aggression and violence have finally been put to a halt through the enforcement of 'Solution Three': the implementation and strict endorsement of the Code. Evocative of genetic, legal and biblical principles, the Code concurrently inscribes the laws of human biology and society. The primacy of genetic engineering is paired off with a social system that propagates the segregation of reproduction and sex. To ban violence, often stemming from inequality and difference between the sexes and races, mandatory homosexuality has been imposed as the norm, and reproduction is strictly regulated through genetic engineering. All children born in this state are cloned from the original He and She – the black man and the white woman whose political activism and empathy for the disenfranchised liberated people from the Aggression Phase. To ensure a pure and non-violent population, all newborns are carbon copies of the original Adam and Eve: 'the baby boys were all brown, just as the baby girls were all white' (30). They are gestated and delivered by Clone Mums, who are carefully selected to bare these children that are not genetically their own.

The Code is not legally prescribed, but functions through a Foucaultian normative apparatus. The ideology of sameness (homosexuality) only marginally tolerates difference (heterosexuality). Despite stringent implementation of the Code, however, two female Council members, Jussie and Mutumba, witness a growing number of transgressions. A couple of 'Professorials' – remnants of the old phase, mostly academics and scientists who refused to adjust to the new norms – falsify the universal validity of the Code on both the biological and social level. As plant geneticists, they observe that a genetically engineered strain of wheat can be wiped out by a single virus. Single crop cultivation endangers the survival of the species. Plant variety is also a trope for social variety. In their private lives, the plant geneticists indulge in a heterosexual love affair, resulting in the birth of two illegal 'natural' children. Besides these two deviants, Jussie and Mutumba are confronted with a rebel Clone Mum, who, in spite of social conditioning and mental inculcation, has bonded with the foetus she is carrying and which is not genetically her own. When, on top of that, they find out that a clone girl and clone boy have fallen in love with each other, the Council members seriously start to reconsider the infallibility of the Code.

Ultimately, the Council is convinced by the irrefutable evidence these deviants provide: apparantly, the Code cannot be applied universally. Moreover, allowing 'difference' and 'diversity' – even though potentially fraught with renewed danger of violence and aggression – may save the human race in the long haul. The Code, as Mutumba notices, is after all an interpretation of the Original's words, and any interpretation is contextually bound. A critique of orthodox bible interpretation is more than implicit: *Solution Three* is an argument against biological determinism as well as against social determinism. Normative heterosexuality based on the biological capacity of a heterosexual couple to produce a child is as nonsensical as normative homosexuality based on the technical possibility of cloning. The State's rule to favour genetics over gestation in fact mirrors the rigidity of Christian morale: the adoration of a Virgin Mother who unwittingly received divine genetic material, yet who is venerated for her gestating quality. Mitchison does not believe in either genetics *or* gestation, or, for that matter, in nature *or* nurture. Instead, she argues the plausibility of complex interaction between genes and environment, nature and nurture. Genetic engineering, even if effacing racism, sexism and violence, should not substitute another indoctrination based on scientific rationalism. At the end of the novel, Mutumba and Jussie announce the implementation of Solution Four: not the elimination of cloning, but engineering a better species by breeding variety.

Mitchison's futuristic novel positions her between Wilhelm's dystopian concept of technological determinism and Piercy's vindication of nurture over nature. She rejects the feminist view that total segregation of reproduction from female sexuality may lead to the degradation of women to gestating birth-machines, or that genetic engineering and reproductive technologies may make female reproductive organs virtually obsolete. But she also resists Piercy's and Firestone's contention that advanced gene-technology intrinsically facilitates the erasure of sexism, racism and (sexual) aggression. Instead, Mitchison argues that science and society incessantly shape each other. Scientific dogmas can never account for social arrangements, as social arrangements can never completely justify a rigid implementation of technology. Any 'code', whether genetic or social, becomes suspect if dogmatically applied. *Solution Three* is a subtle critique of both the scientific 'dogma' prescribing genetic determinism, as any feminist 'dogma' prescribing social determinism. Implied in its title is a rejection of all dualisms that prefabricate our imagination of science and society.

Wilhelm's, Piercy's and Mitchison's science fiction novels can be viewed as responses to the monopolization of genetic knowledge by scientific experts. Like environmentalists and ethicists, feminists attempted to negotiate an expansive definition of genetic engineering. Environmentalists forced scientists to face the potential dangers and ecological consequences of DNA-experiments in the lab; bioethicists challenged them to address social and ethical implications of the new biology; and feminists called attention to the relation between science and society, nature and nurture, and genes and gender. The three contenders that entered the theatre of genetics in the 1970s had one common intention: to redefine genetics as a *social* and *political* issue, and symbolically to 'open up' the laboratory to public scrutiny. At this stage of the debate we saw competing images of genes and geneticists emerge. Environmentalists launched the image of the gene as an autonomous, potentially dangerous bug, an image vehemently denied by scientific experts. And the stereotypical image of the sloppy, arrogant lab worker was counterbalanced by the responsible, ethically considerate scientist.

From the perspective of staging, we witnessed an intensified struggle over professional authority and autonomy. The exclusive rights of scientists to the interpretation and dissemination of genetic knowledge were not only challenged by political activists, but more subtly by journalists. Part of a general recalibration of professional conventions and routines, journalists insisted on more openness and better information. In addition, they no longer viewed themselves as 'secretaries of science', but assigned themselves a new role as public prosecutor or watchdog. By the same token, scientists demanded more control over the distribution of public information and thus their public image. Press policies and media exposure were no longer odious words in the worlds of science. If scientists and journalists previously at least publicly respected the other's authority and autonomy, the strict demarcations between the professions were now fundamentally redrawn.

Although biologists and molecular biologists clearly 'won' the DNA battle in the late 1970s, and environmentalists virtually disappeared from the scene in the 1980s, the politicization of genetics had a permanent effect on its public perception. Ever since the DNA wars, social and ethical considerations formed a valid and accepted concern in the debate surrounding genetic engineering. Geneticists had come out of their secluded labs and had been forced to face social and ethical implications of their work. While they had to give up their authority over genetic knowledge, they got more power over their popular image in return. In the next 'scene' of the theatre of genetics, journalists increasingly appear as crucial allies in the fight for public recognition.

4 Biobucks and Biomania

THE GENE AS MASTER CONTROLLER

The next scene in the theatre of genetics may be captured most appropriately under the heading of 'industrialization'. Just as the politicization of genetics could only be accounted for in the context of social countermovements and democratization processes, the 'industrialization' of genetics can only be properly understood in the context of a political climate that was conducive to economic expansion and capital gains. Reagan's election to office, as well as his British Conservative counterpart's, heralded an era of privatization and deregulation – a perfect breeding ground for the lift-off of an industry that thrived on speculation and investment. Beginning in 1978, biotechnology emerged as the preferred context for the public negotiation of genetic knowledge. While environmental concerns faded away in the 1980s – safety restrictions on recombinant DNA-research had been completely lifted in 1979 – new and old special interest groups entered the stage to discuss their claims within the reset boundaries of the genetics industry: sociobiologists, feminists, scientist-entrepreneurs and anti-capitalists. To retrace their reconfigurations, it is instructive to analyse the revamped images of genes, genetics and geneticists in popular stories on biotechnology. But to comprehend fully how these changes came about, we will also need to look at the shifting dynamics between professional groups.

Ever since the propagation of the helical model the gene had gained autonomy, first as a central object of research, equal in status to the cell and the organism, later as a self-mutating microbe, capable of causing environmental destruction. Vicarious projections of genes as bugs escaping from a lab via a tuna sandwich at least subconsciously contributed to the idea of the gene as the centre of evolution. These gene images in many ways resembled stereotypical depictions of bacterial or viral threats. Viruses or bacteria have a long history of being represented as the enemy of a healthy body, as they were often cast as the invading army probing the preparedness of the local defence artillery.[1] Yet the configuration of the gene also distinctly differed from conventional images of nasty micro-organisms attacking the body's immune sytem. Genes were pictured not as an external, but as an internal threat – a centripetal force rather than a foreign enemy. In addition, the gene was endowed with intellectual human

powers. As Greg Bear's *Blood Music* illustrated, the environmentalist fear
for 'noocytes' generated the idea that a gene might potentially develop a
mind of its own. The gene, in other words, seemed uniquely capable of
controlling its environment and imposing its will on larger organic units,
from the cell up to the body.

This notion of the superior, all-controlling molecule was picked up and
distributed by sociobiologists in the 1970s. The theory that nucleotide
sequences determine what kind of proteins organisms will make, as they
are part of the machinery that controls the production of those proteins,
was taken up most notably by sociobiologists like E.O. Wilson to raise
the status of the gene further.[2] The concept of a unidirectional flow of
information giving historical primacy to the gene undergirds sociobiologists' claim that the gene is the 'master of molecules', and thus in charge
of the organism's development. In contrast to environmentalists' images
of the gene, sociobiologists interpret the autonomous nature of the gene as
a positive, rather than a negative quality. Whereas environmentalists
viewed the 'bug' as something that can easily get out of control, sociobiologists imagined the gene as something that is not only in control of other
organisms, but which is itself quintessentially manipulable.

Of all sociobiology tracts published in the 1970s, Richard Dawkins's
The Selfish Gene has probably been most effective in promoting the gene
as a positive force.[3] Dawkins's self-proclaimed goal in writing his book
was to disseminate his views among a general audience. *The Selfish Gene*
was hailed by the press as a primer for one of the most complicated
scientific theories in twentieth-century biology. A probable key to the
success of the book was the reintroduction of good old factory metaphors
that had surfaced in the 1960s to make the actual working of the gene
sound marvellously translucent. Dawkins embroidered on these images,
but concomitantly attached a distinctly human face to the gene. The
book's title alludes to these anthropomorphic qualities; its introduction
illustrates the peculiar mesh of mechanical and human characteristics:

> The argument of this book is that we are machines created by our genes.
> Like successful Chicago gangsters, our genes have survived...in a
> highly competitive world. This entitles us to expect certain qualities in
> our genes. I shall argue that a predominant quality to be expected in a
> successful gene is ruthless selfishness (2).

The gene, as Dawkins argues, is the fundamental unit of selection, a unit
of heredity endowed with the human characteristic of 'selfishness'. In the
fight for the survival of the fittest, every gene will fight for itself and its
own 'species' as it is not very likely to develop altruistic features. In line

with Darwin's evolutionary theory, genes act in competition, and only the strongest genes win.

It is precisely the conflation of human and mechanical features that accounts for the persuasiveness of these metaphors. Cells, organs and human bodies are reduced to containers for genes to live in. Human beings, in Dawkins's view, are 'survival machines that preserve genes' (19). The cell and the body are completely subordinate to the gene's self-replicating process. Dawkins prefers to think of the body as a 'colony of genes', and of the cell as a 'convenient working unit for the chemical industries of genes'. The gene, on the one hand, has the capacity to replicate itself eternally, eliciting the image of a sheer inexhaustible natural resource – a goldmine or an oil well. On the other hand, it is endowed with unique human qualities – cleverness and selfishness – which make it superior to other molecules and organisms. Since the only purpose of DNA is to survive, the gene necessarily has to co-operate with other genes. Without teamwork and co-operation, the individual gene will not get very far. So the gene, in its hybrid human-mechanic conception, is not only a worker or producer, but also a leader or production manager. The merger of human and mechanical traits is self-explanatory in a capitalist economy where machines, workers and managers are viewed equally as instruments of exploitation. To imagine the gene as a self-replicating control unit within a physical production environment does not require a lot of fantasy from a general audience whose everyday context is made up of commonsense economic laws.

Dawkins's introduction of the selfish gene does more than simply revamp the root metaphor of the factory: it also inserts business and management idioms into what is still basically perceived as an object of scientific research. Genes are not ordinary workers or production units, but they are the 'brains', the command centre that controls the structure and development of living organisms. According to Dawkins, genes control the behaviour of their survival machines – human bodies – not by pulling on puppet strings, but indirectly, like a 'computer programmer'. All genes can do is to program the body beforehand; once its development is under way, they can only sit passively inside. Genes have to 'think' and perform in advance, by '*predicting* and *speculating*' (52). These two words will turn out to be key-terms in the development of the biotechnology industry. The vernacular Dawkins uses to elucidate the working of the master molecule is quite revealing:

Every decision that a survival machine takes is a *gamble*, and it is the *business* of genes to program brains in advance so that on average

they *take decisions* that pay off. ... Human brains (organisms) have
emancipated from their *masters* (genes) by mastering the *technique of
predicting*. ... By dictating the way survival machines and their nervous
systems are built, genes *exert ultimate power* over behaviour. But the
moment-to-moment decisions about what to do next are taken by the
nervous system. Genes are primary *policy-makers*; brains are *executives*
(59–60, emphases added).

According to the Dawkins model, the way in which genes act is analogous
to the way in which businesses are run. Like business, the genetic make-up
is ruled by policy-makers and executives. Genetic laws resemble financial
laws, and the mechanism of prediction and speculation not only prevails in
the stockmarket but equally dominates our genetic disposition. Dawkins
extends the analogy even further by insisting that genes can be thought of
as 'insurance underwriters': the currency used in the casino of evolution is
survival or 'reasonable approximation' (95). These images convert the language of genetics to the language of business. Market analogies, as we
will see later in this chapter, are not at all arbitrary; they tie genetics in
with business, facilitating the birth of biotechnology.

Dawkins is keenly aware of the ideological impact that this introduction
of metaphors has on the public's understanding of genetics. Despite his
assurance that 'we shall keep a skeptical eye on our metaphors to make
sure they can be translated back into gene language' (45), his entire argument exists exclusively by the virtue of these metaphors. And his preten-
sions in stating this theory exceed the goal of merely illustrating an
abstract scientific paradigm: sociobiology appears to provide an all-
encompassing rationale for both natural and cultural phenomena. In one of
the last chapters of the book, Dawkins explains the parallels between
nature and culture. Just as the selfish gene is a survival machine in the
jungle of nature, certain cultural phenomena – 'memes' as he calls them –
have transmuted to survive in the jungle of culture. Strong 'memes' 'naturally' survive in the canons of Western civilization: Socrates and
Leonardo, Copernicus and Christianity all produced the cultural equivalents of genes that have been able to stay afloat in the 'meme pool'. By
means of analogy, the 'natural' operation of genes is projected onto
culture, accounting for the inevitability of culture *as it is*, and defending
the hegemony of certain cultural phenomena by pointing at their 'inherent'
quality – their ability to reproduce and perpetuate themselves in the minds
of generations of people. The meme telos reveals the intrinsic after-the-
fact tautology that comprises the idea of natural selection: those things
that survive do so because they are the fittest, but the way we define 'fit' is

through survival. By means of inference, it is the 'natural' way of defending social and cultural canons.

The idea of the gene as a self-replicating factory, interested only in its own survival and self-consciously programming its incubators to upgrade their chances in the evolutionary race for survival, was eagerly reappropriated and elaborated upon in other fields, such as business, biotechnology and particularly the media. An article in *Newsweek* (1979) enthusiastically recounts how the secrets of the human cell are rapidly being penetrated by biologists.[4] Details of the gene-splicing process are explained in the simplest possible terms: recombinant DNA is the process of 'taking genes from one organism and splicing them into the genes of another'. The journalist reiterates a scientist's hope that they may soon be able to cultivate 'enough man-made bugs to harvest chemical compounds in large quantities'. Although the word 'bug' is still vaguely reminiscent of environmental anxieties, the term also instils the concept of a self-replicating mechanism capable of independently generating products. Small wonder, the reporter concludes, that venture capitalists smell biological magic: the gene has the ability to serve as the equivalent of a mine, if only manipulated properly.

This 'harvesting' or 'managing' of engineered DNA-strings is the focus of a *Newsweek* story, published in 1980, on the new wonders of bioengineering.[5] DNA-splicing is likened to an industrial process: 'Someday, bacteria will be turned into living factories; they will churn out vast quantities of vital medical substances, including serums and vaccines, to fight diseases ranging from hepatitis to cancer and the common cold.' The complex mechanisms of the cellular organism, explained in terms of management and production, shift the emphasis from the gene as an object of research to the gene as an object of manufacture: it may produce useful medicine. This image also concretizes the highly abstract and somewhat scary notion of gene-splicing, turning it into a management process. More importantly, the extended business metaphor generates a web of related meanings, which allows for the gene to be equated with other 'innovative products' – the automobile, electricity and, most recently, the computer. Invoking the early stages of other industrial revolutions, the article creates a sense of excitement, the promise of a lucrative new consumer market. Recombinant DNA-research, as *Newsweek* reassures its readers in 1980, is far beyond the 'Model-T stage'. *Fortune* tags a new biotechnology company as 'the IBM of genetics' and *Time* magazine touts Genentech as 'the next Polaroid or Xerox'.[6]

The reinvigorated factory and business analogies paved the way for the production and marketing of an innovative technology. Starting in 1978, a whole new industry gratefully capitalized on the language of business and commonsense economics. The image of the gene as a controllable controller or a manipulable manager nursed new plots of genetics as a potentially profitable investment. Tales of speculations and lucrative investments countered previously popular stories of bugs escaping into the environment.

BIOTECHNOLOGY AND GENETIC SPECULATION

Biotechnology introduced a new process of manufacturing medicaments and therapies, yielding high expectations for revolutionizing the billion dollar pharmaceutical industry.[7] The very word 'biotechnology' – a word invented on Wall Street in the late 1970s – reflects the amalgamation of science and commerce; it transfers pure (bio)science into a set of techniques or tools, which can be used to generate actual products.[8] Since the hundreds of new biotechnology companies that started in the early 1980s did not have any substantial product to show until 1985, investors and the public had to rely to a large extent on images – images that helped them imagine what a gene was and what genetics could do. New plots of genetics as a potential 'gold rush' were needed to create and sustain the prospect of a burgeoning new field with unlimited potentials for new markets and lucrative applications. Two terms introduced by Dawkins were crucial in framing popular biotechnology stories: 'prediction' and 'speculation' appeared to be the shared idiom that coagulated scientific excitement with investor's dreams.

In the nineteenth century, the theory of heredity had become prominent mainly because of its prophetic nature. Knowledge of genetic laws did not simply imply the ability to produce 'objective', scientifically induced information concerning someone's health, but it connoted the ability to forecast one's future. This element of speculation proved even more significant in the era of the new genetics, when it came to calculating the exact risks of developing disease already present in the genes. Then as well as now, predictive claims – the calculation of chances for survival – carried an aura of reliability because they were purportedly based on scientific knowledge. Scientifically warranted claims turn speculation into prediction – a shift in emphasis that tends to eliminate uncertain or disputable factors from scientific reasoning. The technical ability to predict a person's future state of health on the basis of present genetic markers creates two new needs: it urges people to find out about their own future chances for a healthy life and procreation, and it raises hopes that the very technology used to determine the causes for defects will also lead to cures. Speculation and prediction thus merge connotatively, engulfing hope and extrapolation in the process.

Speculation is also a common trope in the world of investment and high finance – prophet and profit being more than acoustically related. The stockmarket, a world prospering on prediction, hopes and risks, discovered the speculative value of genetic engineering in 1980, setting off a frenzy in public stock offerings and, at a later stage, buy-outs of small private biotechnology companies by multinational pharmaceutical corporations. The promise of recombinant DNA-technology to develop cheap remedies to heretofore uncurable cancers or congenital diseases led investors to pump considerable amounts of money into an industry that had yet to come up with its first marketable product. During the first five years of 'biomania', companies managed to sell 'gene dreams' purely on the basis of speculation. But a shared preference for speculation cannot solely account for the emergence of a biotechnology boom. What underlying plots made the story of biotechnology look so convincing and lucrative?

The early development of the biotechnology industry can hardly be viewed separately from the creation of a need to predict and cure genetic diseases. Prediction or genetic screening seemed an absolute precondition for both preventing and curing genetic flaws. The image of the gene as a self-replicating mechanism conceptually allowed for notions of 'damage' and 'repair', implying the potential for future technologies that 'fix' genetic flaws. But the 'fixity' of genes necessitated the need for advanced fortune-telling – calculations of the likelihood of contracting serious diseases or reproducing congenital defects. For an industry that had yet to come up with its first product, these images were important conceptual keys. They not only helped imagine what kind of products might emerge from abstract technological processes, but also helped establish the need for curative medicine, whether drugs or gene replacement therapies. Genetics became a story of crippling disease and glimmering hope for prevention and cure.

Not coincidentally, the first real 'products' furnished by the biotechnology industry were diagnostic tools: genetic tests which can be applied in the early stages of pregnancy when abortion is still an option. Produced as a tool-kit to screen for a set of genetic defects, in the public mind these tests became increasingly known as indicators of a child's future health. They were marketed particularly to women who were anxious to assure the health and normalcy of their offspring. Screening tests firmly located genetics in the context of the hospital or the clinic, rather than the laboratory. Popular gene narratives poured out a new medical vocabulary over

their readers: genetic counselling, genetic therapy, genetically based diagnosis, gene pool and other clinical terms entered common parlance.[9] Most
indicative of the 'predictive' nature of this new genetic industry was the
growing presence of genetic counsellors – a group of professional fortune-
tellers whose task it is to calculate chances for carrying over genetic
deficiencies to one's offspring. As a *Newsweek* reporter explained:
'Eventually, counsellors may be able to provide a "genetic profile" of a
person's susceptibility to diseases so preventive measures can be taken.'[10]
The clinical environment in which these tests and counsellors materialized, helped to associate fortune-telling with medicine rather than with
commerce.

But biotechnology did more than create plots of prediction and new
markets for genetic tests and counsellors. Diagnostic tools also trained the
general public to see genetics as the next great step in technological
progress, stimulating the so-called geneticization of the social mind. The
fast implementation of genetic screening tests was symptomatic of an
increasingly pervasive way of thinking about human health and disease in
terms of risk management. Dorothy Nelkin and Laurence Tancredi, in
their book *Dangerous Diagnostics*, describe the impact of genetic tests on
the public mind:

> The increased preoccupation with testing reflects two cultural tendencies
> in American society: the actuarial mind-set, reflected in the prevailing
> approach to problems of potential risk, and the related tendency to
> reduce these problems to biological or medical terms. Actuarial thinking
> is designed to limit liability … The information derived from these tests
> becomes a valuable economic and political resource.[11]

This 'actuarial mind-set' manifests itself abundantly in the vocabulary of
news media articles on genetics and biotechnology. Terms such as
'genetic vulnerability', 'genetic predisposition' and 'genetic tendency'
emphasize the inherent risks of living, and the future vagaries that are
already stored in our genes. It is exactly these uncertainties that enable
marketeers to exploit fears: fears of birth defects, 'problem' pregnancy
and disease. Speculation almost unnoticeably shades into prediction, as
genetic screening becomes a 'probability statement about future risk'.[12]

While diagnostics were not a 'classy' product – they do not promise a
cure – they were pivotal to the creation of a consumer market. The rising
popularity of diagnostics boosted both scientific and market expectations,
and stoked up excitement about future applications – grand hopes for
miracle cures. The worlds of science and finance were equally involved in
this process; their contributions to the view of genetics as the new medical

cornucopia are inextricably intertwined. Expectations of cheaply produceable cures for which there would be a strong demand set off a frenzy on
the stockmarket when companies started to raise capital investment.
Prospects of marketable products – diagnostic tools, medications and gene
replacement therapy – virtually depend on the basic assumption that
organisms function predictably, that they behave according to decodable
scientific formulas. Images of the gene as a manageable and controllable
factory or a self-mutating resource frequently surface in 'biomania
stories'. A resource that works predictably and produces unremittingly
appears a safe object for speculation and investment. Moreover, from 1980
onwards, engineered genetic strings, like other industrial products and
processes, could be patented and thus 'owned', procuring exclusivity in
future mass production.[13]

As Robert Teitelman has meticulously documented in his case-study of
Genetic Systems – one of the small biotech firms that had started in the
early 1980s – 'biomania' was fuelled by inflated expectations of finding
magic bullets: effective cures which exclusively target cancerous cells
without producing any side-effects. The whipped up dream of interferon
as the ultimate cure for cancer, according to Teitelman, was the 'dress
rehearsal for the biotech industry'; after the interferon hype, many similar
promises, such as interleukins, hit the publicity mill.[14] Besides the promise
of cures, other prospects of future products that are frequently mentioned
in the press are bacteria that 'eat' oil spills, turn garbage into alcohol, and
transform methane into protein suitable for animal and human consumption. These product-images seemed to counterbalance the still lingering
images of engineered organisms escaping from the lab. Projected biotechnology products bolstered the idea that they will save the environment
rather than harm it.

With hindsight, it is quite remarkable to see how easily capital investors
bought into the complex language of cells, monoclonal antibodies and
double helixes. It may well have been the mysterious intricacy of molecular theory that strengthened the faith of investors, but it was certainly the
new industrial and commercial image of genetics that justified their belief.
As Teitelman explains:

> Talk of medical breakthroughs became pretexts to raise more money.
> Capital accumulation was confused with speculation; rhetoric was mis-
> taken for reality. Companies wielded complexity like a weapon; results
> so sketchy and ambiguous that they could be interpreted freely, fantasti-
> cally. The sheer distance from lab to clinic, from cell culture to human
> patient, created a sort of *imaginative space* and nurtured dreams of mira-
> cles and money (214, emphasis added).

If we compare the 'golden' image of genetics with previously dominant notions of moral fear and environmental anxiety, one common denominator appears to underlie the creation of new plots: the role of 'imaginative space'. Imagination is required to translate abstract scientific models and principles into commercial applications; the distance between the laboratory and the clinic is bridged by scientists and investors who are able to 'design' needs and products. Just as fanciful scenarios of escaping bugs were used to stimulate the generation of safety guidelines and policies, fantasies of gold-crunching genes were now deployed to wheedle money out of private investors. Story-lines of cheaply produceable medicine that would help prevent a number of serious diseases, tool-kits that enabled prediction and thus prevention, and the glimmering hope of curative gene therapy paved the way for a 'genetic mind-set'.

Along with the adjusted images of genes and the imagination of genetic plots came new images of the geneticist. Far from the 'immoral scientists' or the 'sloppy lab worker', the character of the geneticist metamorphosed first into a crystal-gazer and subsequently into a hybrid scientist-entrepreneur. The cover of a *Newsweek* story (1984) on the state of the art in biotechnology shows a man in a white coat peering into a glass ball filled with a mysterious substance.[15] His hands seem busy exorcizing the glass ball – arcane symbol of speculation. The figure of the scientist represents a variation on the sorcerer's apprentice – a high-tech fortune-teller who is able to predict someone's likely genetic future. The reader is no longer confronted with 'gene splicers' or impersonal lab technicians explaining the intricacies of the human cell, but with doctors in white coats. The article recounts the story of a desperate couple that wants a healthy baby, yet who are themselves at risk for sickle-cell anaemia; they 'carry a genetic time-bomb', according to *Newsweek*. The work of 'gene-doctors' has enormous implications for the diagnosis and treatment of diseases since, as the reporter claims, 'there is no disease that does not have a genetic component'. Later stories increasingly featured the geneticist as a succesful hybrid of scientist and entrepreneur, someone with a face, a name and a white coat.[16] Emerging from relative obscurity, genetic engineers became conspicuous players in the field. The viability of start-up biotech companies was largely judged by the names and fames of their affiliated scientists, so their popular images were an important asset in the drive to raise private funds. In an industry without products, the persona of the scientist became an important asset.

Popular images, rather than being a reflection of the public's hopes or anxieties, had previously played a role in the negotiation of strict regulation policies, and now became a factor in the promotion of a new industry.

For one thing, the new language enabled public negotiation of genetics in terms of industry and business rather than just in terms of moral, ethical or environmental anxiety. Genetic research had to be 'sold' to investors and the general audience largely on the appeal of its products, the plausibility of its plots and the reputation of its scientists. The industry had become, so to speak, 'image-dependent' and the steering of this vulnerable image would receive the highest priority. But before we look at how scientists and journalists together navigated biotechnology's images, we first need to look at the foremost antagonists of the 'gene business' to see how they were scripted in the theatre of genetics.

BIOTECHNOLOGY BUSTERS

The impact of sociobiologist vernacular on popular configurations did not escape the attention of detractors of the new industrial genetics. Sociobiologist theories and the subsequent prioritization of industrially based biotechnology have been severely criticized by both biologists and feminists.[17] Richard Lewontin has pointed out how sociobiology is nothing less than a genetic defence of the free market, *laissez-faire* capitalism operative in the realm of genes. He observes an intimate connection between biology and ideology, explaining that sociobiologist theory can flourish only in a capitalist society in which the view of the individual as an autonomous social atom is 'naturally' matched by a reductionist view of the human body as a survival machine. One of the most effective weapons in the advance of new genetic theories, Lewontin argues, is the insistence on genes as active agents that are operating totally 'inside us', regulating and defining the human species, whereas the environment is configured as a passive context, existing outside us. Concepts like 'master molecules' and 'survival machines' give rise to the false dichotomy between nature and nurture, a postulate that goes back to Darwin. 'The new biology has become completely committed to the view that organisms are nothing but the battle grounds between the outside forces and the inside forces' (83). Industrial and business metaphors, such as the factory and the business manager, help substantiate the claim that the gene is ontologically prior to the individual as the individual is to society. In Lewontin's view, sociobiology reflects the ideology of neo-conservative libertarianism, contending that society is best served by each individual acting in its own best interest.

Strong objections to the industrialization of genetics also originated from a particular group of feminists, who organized themselves in 1984 under the acronym FINNRAGE – standing for Feminist International Network of Resistance to Reproductive and Genetic Engineering. FINNRAGE feminists fulminated against the commercialization of genetics, which combined the evils of a capitalist and patriarchal society. According to FINNRAGE spokeswoman Renate D. Klein, geneticists have embarked on a mission to turn women into machines in order to dissect the reproductive process into marketable parts, like genes, eggs, wombs: 'There will almost certainly be domination, appropriation and exploitation of women to a yet unprecedented degree, unless women organise to fight back.'[18] Society is simply divided into two self-explanatory categories: male scientists and female victims of technology. Geneticists' ultimate goal, as Klein tries to convince her audience, is to create artificial wombs that render women's bodies completely obsolete, and allow 'them' to manipulate genetic material endlessly. Women who do not share her position on these issues are declared out of touch with reality. They believe in fairy tales that men have inculcated into their minds: 'Some women unfortunately have been socialised into imagining that if they revere the clone kings, ... they will one day perhaps become the queens accompanying the kings' (265). Gena Corea, another prominent FINNRAGE member, further amplifies the cluster of industrial images by explaining how the 'pharmacrats' will take over the genetic industry: 'With their laboratories and machines, men will produce more "perfect" babies than these women will produce with their fleshy, natural bodies.'[19] Eventually, geneticists will try to establish a 'genocracy' supported by Western capitalism, that justifies the exploitation of women's bodies as natural resources. From a Marxist–feminist viewpoint, genetics is assessed in terms of domination and control, exploitation of resources and assembly line reconstruction of reproductive parts.

Yet unlike Lewontin, who meticulously dissects the binary and hierarchical image-systems of industrialized genetics, Corea and Klein criticize the market mind of genetics, using exactly the same terminology and imagery favoured by *Business Week* and other magazines as they talk about 'gene kings' and 'managed resources'. The images that FINNRAGE-adepts deploy to prove the inherent threat that genetic engineering poses to female reproductive power supplement the ones promoted by proponents of the biotech industry and endorsed by mainstream media. Feminists seem to incorporate the very same rhetoric to script themselves as victims of the male urge to manage and control nature with technology. Although radically opposite in its intentions, the philosophy of FINNRAGE is to a large extent derived from the same essentialist underpinnings as sociobiologist theory. The assumed split between science and nature accounts for the configuration of genes as production machines and people as environment in which genes can thrive and survive. Similarly, images of scientists exploiting genetic material as reproductive resources and wombs as gestating environments are based on the contention that technology and nature are inherently adversarial categories.[20] As feminist critics have observed, FINNRAGE feminism is a peculiar mirror image of the patriarchal ideology it opposes, or, as Hillary Rose has stated: 'it frequently reduces women (and men) to nothing but biology, in which, in this particular struggle, reproduction is the central function.'[21] Despite their apparent antagonism, the reasoning of sociobiologists and FINNRAGE-adepts is perfectly commensurate in that both arguments deploy a logic rooted in the ostensible split between genes and environment, nature and culture.

Whereas sociobiologists gave leverage to the image of the gene as a micro-factory and the organism as its environment, Marxist-feminists responded by promoting the image of the female body as a 'natural' environment for genetic experiments, and reproductive functions as a process managed by scientists. Their imagery and binary frameworks reflect an assumed opposition between the untainted, 'natural' bodies of women and the exploitative, infested realm of industrial science – science corrupted by the interests of capital and patriarchy. A mirror or complement of hegemonic industrial images reinforces rather than subverts the hierarchy.

FUSING TWO CULTURES

Feminists are not alone in holding on to the bipolar framings to which they so fiercely object. Popular stories on the gene industry reflect a general tendency to stick to oppositional frameworks. Biotechnology is often referred to as the epigone of an ultimate 'fusion' between two cultures: the world of science and the world of business. That science or university-based research was untainted by business interests until the early 1980s is a common misperception. Academia and commerce have always been interrelated to some extent, and the desire for that pristine moment in the not too distant past when these two domains were completely separate is unwarranted nostalgia. It is a truism, however, that with the rise of start-up biotechnology firms the interests of both domains got directly and completely intertwined. Yet in staging the wedding between these two cultures, the role of a third party is often seriously underestimated: the media played a significant part in the courting process of two presumably unlikely partners.

One of the most sensational engagements between genetics and high finance was the public stock offering of Genentech in 1980. This small biotech firm, established in 1976 by scientist Herbert Boyer of the University of California at San Francisco and venture capitalist Robert Swanson, was the first to raise private money through the stockmarket, in order to increase working capital. The announcement of the public stock offering caused extreme agitation in the press. Almost every popular and professional magazine, from *Business Week* to *Science Digest*, paid attention to the forthcoming event. Favourable publicity caused expectations to rise: 'It will be only a few years until microbal factories will be competing with conventional processes in markets worth millions of dollars annually', *Business Week* predicted; and 'investors are queuing up to buy what some believe will be one of the strongest new issues of the 1980s', according to *Time*.[22] In the weeks before the event, articles in the press featured headlines such as 'The Miracles of Spliced Genes' (*Newsweek*), 'The Hunt for Plays in Biotechnology' (*Business Week*) and 'Cloning Gold Rush Turns Basic Research into Big Business' (*Science*). Genentech's 'coming out' deal was delayed by the Securities and Exchange Commission because of strongly favourable press coverage, yet nothing could prevent an immediate excitement. Within hours after the stocks were offered at $35 a share, the price skyrocketed to $89 – an all-time record for an initial public offering. In the days after the stock offering, magazines and newspapers profiled young postdoctorates and established scientists who became millionaires overnight, because their salaries had previously consisted of worthless shares of their employers' firm.[23] Many other small start-up companies profited from the goldfever triggered by Genentech's success, which set off a 'gold rush': within a few years, over one hundred new biotech companies were established, raising some $500 million in private capital.[24]

The new scientist-entrepreneurs perfectly understood the pivotal role that journalists played in the promotion of their popular image and their reputation. Publicity went beyond the need to reach investors, and small biotech business turned directly to mass media outlets to gain public support for this new kind of industry. Journalists became the prime targets of scientists and entrepreneurs who wanted to announce the results of DNA-research, not so much to satisfy the public's hunger for information, but to boost public expectations. Rather than following the conventional track of scientific communication – first publishing the results in a major scientific journal before allowing the press to report on these findings – gene engineers working for small biotech firms skipped the first part of the process and turned directly to the lay press. Several months before its stock offering, Genentech researchers called a press conference to announce the discovery of a new recombinant DNA-technique that produced human insulin. The story became front-page news everywhere, but the actual scientific data were not publicly available until three months later. Claims made by Genentech researchers turned out to be entirely provisional; however, nothing could undo the boost in Genentech's image – a welcome asset right before its public stock offering. In a similar vein, on 16 January 1980 a group of scientists affiliated with Biogen announced to the press its success in the cloning of interferon-producing bacteria. After the press conference, journalists were presented a draft of a paper prepared for submission to the *Proceedings of the National Academy of Science*, the accuracy of which no scientist had had an opportunity to verify. Needless to say, Biogen gained many points after the public announcement.

In an editorial, the *New England Journal of Medicine* (*NEJM*) expressed its strong disapproval of this unprecedented phenomenon which it aptly labelled 'gene cloning by press conference'.[25] The *NEJM* sharply condemned the new practice of short-cut science communication, the need for which it ascribed to increasing competitiveness and mounting pressure to win patents: 'Their [scientists'] behavior does not contribute to either good science or good science reporting, and it is incumbent on the scientific community as well as journalists to debate this issue widely' (745). In one of its later issues, the *NEJM* warned that the new industrial connection of DNA-research is 'bound to impinge on the public's impression of the scientific enterprise', since industrial support foregrounds marketability and profitability. In the long run, so they claim, the 'good name' of the scientist will pay off.[26]

That 'good name' obviously depended on an unconditional adherence to academic rather than commercial values, such as scientific integrity and a dedication to scientific methods and goals. The professional group that had been put in charge of biotechnology's public recognition and reputation understood very well which values to sell to what group. To investors and stockholders, public relations officials emphasized the commercial potential of prospective products and looming patents. Early prospectuses of biotechnology companies show an overwhelming preference for the language of business and industry. Genentech boasted that its potential returns would be measured in 'dollars per milligram instead of cents per pound', and though its stock might be 'highly volatile' this also implies that you cannot expect to win unless you are willing to take big chances.[27] Cetus, another start-up biotech company that went public shortly after Genentech, put in its first annual report that 'microorganisms are "factories" producing useful products'.[28] The image of the micro-factory, introduced by Dawkins as a metaphor, was actually taken very literally by marketing experts in the biotech industry. It certainly helped investors 'imagine' the potential for profit. These metaphors were in turn adapted by the media, as they hailed the industrial exploitation of genes in a variety of popular magazines, from *Fortune* to *People*.[29]

When addressing a general audience, however, public relations managers strictly emphasized academic values, and played down the economic interests of biotech firms. Via advertisements, they promoted their companies' intrinsically humane, clinical goals and intentions, banking upon the name and fame of the company. The issue of the *New Scientist* commemorating the fortieth anniversary of the Cambridge Institute for the Molecular Structure of Biological Systems (1987) provides a case in point.[30] Eleven articles, each featuring a remarkable advance in molecular biology in the past decades, are interspersed with full-colour advertisements for biotechnology companies. In that for Wellcome Biotech, for instance, an enlarged model of the double helix is accompanied by phrases that extol the company's inherently humanistic intentions: 'Today, drawing on the scientific achievements of many of the world's leading academic researchers, Wellcome Biotech is constantly developing new techniques for treatment and prevention of disease.' The ad culminates in unabated praise of the firm's research team, 'men and women working at the farthest reach of scientific discovery'. The corporate image seems based upon the devotion of its employees to curing disease by developing recombinant DNA-vaccines: 'Their commitment is an inspiration to us. Our commitment is to the continuing race against disease'. Another ad, for British Biotechnology, displays a large picture of the young Watson and Crick in front of their double helical scale model, aligning their theoretical achievements with an impending new era of discoveries in biotech products: 'As a new innovative company we are at the forefront of medical research and development.' Most advertisements for biotech firms underscore the collaboration of their industrial research teams with academic research centres or groups. Industrial research and development is explicitly labelled as 'independent' and is said to be executed in co-operation with the world's 'leading *academic* researchers'.

Despite the obvious fusion between academia and commerce, the hierarchical order between the two domains remained inscribed in the public consciousness. To ameliorate the 'corporate' image of biotech companies, public relations people paradoxically appeal to 'academic' values. The hierarchical dichotomy is vigorously endorsed by the very scientist-entrepreneurs who instigated the fusion. Apparently, the strict demarcation between science and commerce also forms a fixed 'script' for evaluating the new roles of hybrid actors that increasingly populate the stage. Leading faculties in molecular biology and genetics obtained lucrative consulting arrangements and industry positions.[31] By the end of the 1980s, there was hardly any respected molecular biologist or geneticist left at a university who was not somehow involved in the biotechnology industry, either as a consultant on the Scientific Advisory Board, a board member or an executive.[32] Biologists were offered handsome salaries upon transfer from universities to for-profit firms. And in a society where money-making is highly respectable, it should come as no surprise that scientists were lured by the prospects of higher salaries and stock dividends. Geneticists themselves in fact became commodities, as universities attempted to avoid an exodus of qualified personnel by giving in to their pecuniary demands.[33] The result of the emerging hybrid scientist-turned-businessman, according to *Science Digest*, was a 'culture chasm' in the biotechnology industry – a fissure between scientists and managers who 'inhabit quite different cultures where different methods of communication and perceptions of time, as well as styles of work and even dress, can be distinctly different' (36).[34]

Although the geneticist was now hailed in the media as both an objective scientist and a smart businessman, his direct financial involvement did not seem to compromise his image as a selfless pursuer of knowledge. On the contrary, the cosmetic split between industrial and academic workers was often exploited to emphasize the trustworthiness of the scientist. In fact, geneticists or molecular biologists who hold positions at universities are important qualitative assets to a company. Not coincidentally, a number of scientists who left academia, often to start their own companies, resumed their academic positions at a later stage because the status and reputation of university professors turned out to add significantly to their market value as researchers for profit-making institutions. The merger of science and industry seemed paradoxically attended by a persistent symbolic separation between the two domains of academia and commerce.

Whereas in this model the individual scientist was allowed to engage moderately in commercial activities – lured by financial gain – academia itself was commonly held up as the classic example of an institution untainted by commercial interests. But commercialization of university genetics research became the rule rather than the exception. The push for privatization encouraged universities, in the face of considerable budget cuts, to look for mutually profitable exchange of resources; the burgeoning biotechnology industry provided a unique opportunity to explore yet uncultivated terrains of co-operation. The loom of possible high returns on investments in expensive lab equipment proved more than a little tempting to 'cash-starved colleges'. The first actual attempt of a university to initi-ate its own enterprise, however, caused a furore among academics and in the press. When Derek Bok, President of Harvard University, announced plans, in November 1980, to start up a genetic engineering firm in which the university would be a minority stockholder, its faculty firmly opposed the plan. The potential for conflicts of interest appeared too real and the academic community insisted on the university's unconditional autonomy. Press reactions to Bok's proposal reflected the general outcry against com-mercialization, when it talked about 'Ivy League capitalism' (*Washington Post*) and 'Harvard Inc'. (*Time*).

This storm of criticism sent a strong signal to political and academic leadership, resulting in a series of congressional hearings to explore the effects of commercialized biomedical research on the academic research climate.[35] Reminiscent of the Berg group who organized a conference to discuss safety regulations in biolabs in 1974, Stanford's President Donald Kennedy convened a meeting in Pajaro Dunes, California, to discuss public and institutional policies in a growing grey area between science and commerce.[36] In contrast to the Asilomar conference, no applicable set of guidelines or principles emerged from this gathering; although the five big university players in the biotech industry (Stanford, Harvard, MIT, University of California, and California Institute of Technology) admitted the need for such 'rules of conduct', they came up with only a rough draft, containing no more than rudimentary recommendations and suggestions.[37] And, unlike the effort the Berg group had invested in regulating press attendance of the Asilomar conference, at Pajaro Dunes no reporters were allowed to follow the discussion. Not unexpectedly, the press reacted sceptically to the intentions of universities to stimulate closer ties with the industry. Between the Asilomar meeting of 1974 and the Pajaro Dunes conference of 1981, the agenda had changed from how to handle environ-mental hazards to how to manage the impact of commercial pressures on university research. And predictably, as Edward Yoxen observes, 'many of the same people who had said that the risk question could be handled were also of the view that commercialisation was really no problem either, as long as there were some rules of conduct and everyone agreed to stick to them'.[38]

Although Harvard was forced to refrain from its plans to form a for-profit gene-splicing company, they did accept large gifts from the chemical and pharmaceutical industry to fund its university lab and medical school, to expand research into genetics. Universities were rather creative in inventing legal constructions which permitted sponsored research and returns on investments.[39] The various forms of co-operation between universities and companies ranged from research grants and contract research to industrial affiliate programmes, co-operative research centres and long-term contracts between one university and one company.[40] These hybrid constructions were the direct result of the ambivalent recognition that, while money injections to cash-draining bioresearch departments were inevitable, a university's outright shareholding in a company was undesirable.

The media consequently framed academia and business in oppositional roles, and positioned themselves either for or against commercialization of genetics. Reactions in the popular press reflect a strong initial public reticence to accept the evolving courtship between science and commerce. As *Science Digest* warns in 1982: 'The lure of profits is changing the face of the biological sciences in an explosive manner. Just six years ago there was no genetic engineering industry. Today scores of companies have joined the fray and new competitors are coming in all the time.'[41] Most vehement objections against commmercialization, of course, came from the liberal press. Biologists, *The Nation* laments in 1979, used to be innocent scientists, but now that they have entered the marketplace they have to play by rules of business.[42] Scientific results and research information can no longer be freely exchanged in a competitive environment where new discoveries may result in exclusive patents. Pressured by investors' expectations, scientists may be forced to produce substandard or fraudulous claims, just to satisfy shareholders. And above all, the prospect of research agendas being dictated by commercial interests rather than public needs, set off a jeremiad against the pressures of 'biobucks'.

Ten years later, *The Nation* (1989) observes that many of these predictions have become common practice.[43] Under the new Technology Transfer Act, universities and publicly funded government agencies are allowed to co-operate with private industries, and individual scientists can earn limited supplementary income from commercial activities. In order to fund expensive but ground-breaking experiments in gene transfer therapy, a team of National Institutes of Health (NIH) scientists signed a co-operative research agreement with Genetic Therapy Inc., a small biotech company in which both the NIH and French Anderson, the leading scientist of the NIH team, are shareholders. These researchers held back crucial information during an early phase of the approval process for the transplant experiment, to protect their data from competitors.[44] In the eyes of *The Nation*, these commercial trade-offs seriously undermine the public interest, and the article ends in an ominous warning:

Inevitably, Mary Shelley's tale of science gone awry enters the gene therapy controversy. ... French Anderson and his team ... are tinkering with the very essence of life. To do so at all is hazardous enough. To do so in an atmosphere increasingly marked by greed invites calamity (478).

The 'incorporation' of bioscience was happening at such a pace that it was soon impossible to tell science and commerce apart, as their collective spiralling around the same genome seemed only 'natural'. Commercial market values, as critics Hill and Turpin contend, penetrated the public science sector to such an extent that it was 'not merely a pendulum that has swung too far but instead the commercial marketplace is sitting inside the processes that forge the global constitution of society's knowledge'.[45] The remarkable '*Anschluss* of academic territory with the corporate goals it now served' was either sharply condemned by anti-capitalists and the liberal press, or it was favourably endorsed in the mainstream press, yet always departed from an explicit insistence on a possible distinction between science and commerce.[46] Despite the obvious fusion, genetics was overwhelmingly scripted in terms of distinct oppositional and hierarchical categories.

The press, rather than being an innocent bystander at the wedding of two cultures, had become a constitutive factor in a *ménage à trois*. Although they firmly endorsed the constitutional split between science and industry, they were themselves caught up in the mangle between institutional values and commercial pressures. Their interest in biotech stocks as 'hot commodities' had driven them to team up with biotech-entrepreneurs to help establish a favourable popular image. Scientist-entrepreneurs, for their part, learned to seduce journalists with scoops, and thus co-opted some of the journalist's power over the dissemination of scientific knowledge. The press had now gained considerable access to what was once ivory tower science, but had to yield power in return. Public relations managers appeared prime shapers of popular knowledge, as they increasingly controlled the flow of information between scientists and the general public. Yet despite various tangible coalitions between the worlds of academia, commerce and journalism, there seems a remarkable insistence on the demarcation of each respective professional authority.

THE GENETIC UNDERWORLD

Besides in the scripting and staging of biotechnology stories, the 'industrial' context of genetics can be inferred from the preferred settings of popular stories. Obviously, the growth of the biotech industry came along with a change in physical setting from the laboratory to industrial compounds. But the shift can also be traced in the symbolic and imaginary allocations of genetics either to the 'underworld' or the vast worlds of unexplored space. Both invocations seem peculiar retakes on previously introduced spatial settings.

The coalescence of industry and academia is typically reflected in the geographical distribution of the hundreds of new biotech companies established in the early 1980s. University 'science parks' emerged at sites close to campuses like Stanford and Harvard. The new 'gene factories', as they were also called, were close enough to university premises to build on these ties, and close enough to other businesses to absorb the entrepreneurial spirit. Many of the élite bioscientists enthusiastically endorsed this literal redrawing of the borders between industry and academy, because, although they cherished the academic spirit, they were extremely weary of the many bothersome bureaucratic obstructions that academic life imposed on their research. Besides relieving scientists from the burden of serving on committees, teaching and administration, industrial settings offered additional advantages: better equipment, more resources and a highly motivated team of colleagues.[47] The physical setting of university science parks undoubtedly facilitated the merger of the two cultures, as it became the literal site of boundary negotiations between science and industry.

The emergence of science parks as mixed academic-entrepreneurial arenas also translated into a growing body of literature on biotechnology, creating its own symbolic imagery and mythology.[48] Reminiscent of older, conventional images, genetics was often associated negatively with the 'underworld' – as illustrated in *The Eyes of Heisenberg* – or positively linked to the discovery of new living space. Popular stories on biotechnology reinvigorate these traditional images, and attune them to the realities of 'genetic enterprises'. A journalist for *Esquire* portrays six 'world class scientists', all involved in genetics research, who dedicate their talents and energy to the exploration of new worlds.[49] To them, genetics is more than just a discipline: 'It's a way of looking at things, a way of thinking' (209). They descend into spatial territory that no one before has laid eyes on. The six 'bright young men' are on the brink of discovering a completely new dimension of space:

> When they are on the job, these six are deep in a subcellular realm, a place beyond microscopes where the only light to see by is the light of ideas. Getting there and succeeding there requires a rigorous application of what can only be described as scientific imagination (197).

Molecular biologists, as the reporter explains, live in a world of their own – the world of the infinitessimal – where a 'gene seems as large as a ladder and where cells are the size of ballrooms'. This space odyssey into the infinitessimal echoes the filmic concept of *Fantastic Voyage*, as described in Chapter 2. An indispensable tool on this expedition into the subcellular realm, according to *Esquire*, is the scientists' imagination: their ideas have to enlighten us and stimulate our understanding of this invisible and inconceivable terrain. The very concept of 'the new world', equally appeals to scientists and entrepreneurs: whereas scientists are encouraged to discover previously unimagined spaces, investors and entrepreneurs further develop and exploit these natural resources.

Associations of molecules with vast unexplored space abounds in the many book-length non-fiction accounts of biotechnology adventures that appeared in the 1980s. In addition to glorifying the bold border crossings of scientists going commercial, journalists also embroider on the biotechnology bonanza by abundantly mixing factual description with fictional plotting. Stephen Hall's *Invisible Frontiers*, for instance, recounts the 'race' to synthesize the insulin gene between 1976 and 1982.[50] The book is written as a classic adventure story, even though it is based on interviews with the scientists involved. As its cover claims, the book is 'both an authentic and vivid piece of science history' and a 'stunning, fast-paced thriller'. The framing of this story is clearly reminiscent of Watson's *The Double Helix*. Two teams, one in Harvard led by Walter Gilbert and affiliated with Biogen, and one in San Francisco headed by Herbert Boyer and co-operating with Genentech, compete with each other to isolate the gene for synthesizing insulin. Much more than just the 'discovery' and the honour of winning is at stake in this race. Besides yielding a Nobel Prize and the victory of Boyer's Genentech team, this race symbolizes the very 'birth of the biotechnology revolution', as Hall contends.

Invisible Frontiers not only borrows its narrative frame from *The Double Helix* but also echoes the popular equation of molecular and lunar dimensions. Spatial metaphors further enlarge molecules to the size of the galaxy and beyond:

> Like the astronomers who inferred galaxies and quasars from invisible radio waves, molecular biologists inferred from the almost invisible life processes of bacteria and viruses the constellation of biochemical activities that applied to all living things, to human beings as well as to microscopic bugs (9).

To explore infinitely large space is obviously more heroic than to travel into the inner sanctum of the infinitessimal cell. Yet the imagination of vast unexplored space is consistently compared with ancient mythology. For the discovery of new lands through intergalactic travels, scientific imagination does not suffice; a pioneer mentality is indispensable. The trail-blazers of a new cellular age are presented as rugged individuals who are constantly confronted with new hurdles to overcome. First, there are bothersome safety guidelines that frustrate a smooth and rapid pursuit of experiments. Then there are critics like Jeremy Rifkin whose actions are more than a minor nuisance. And, last but not least, scientists have to train themselves to fight the endless bureaucratic obstructions erected by government agencies and universities.

This last hurdle in particular makes it understandable why scientists are almost *forced* to turn to private investors. Arguing that strict funding regulations and requirements fundamentally hamper freedom of inquiry, the author appeals to the basic instinct of the entrepreneur, the spirit of free enterprise. In Hall's eyes, universities are too monolithic and unwieldy to accommodate these fast developing technologies. The expediency of private investment in technological progress is a trope intrinsic to the ideology of capitalism. While competition between two teams of scientists is presented as 'academic' competition – reminiscent of the team competition Watson introduced in his account – in *Invisible Frontiers* the academic race almost imperceptibly shades into market competition. The ultimate victory does not only result in public fame, such as a Nobel Prize or academic recognition, but more importantly produces patents on manufacturing processes – exclusive claims on the prospective production of medicine.

Invisible Frontiers manages symbolically to relocate genetics research in the frontier of human knowledge. The conventional setting of the new frontier attaches old values to new and previously suspect experimental research. Science and commerce naturally blend in the imagery of explorers' expeditions. New lands can evidently be conquered, owned and exploited. New genes can be discovered, patented and endlessly replicated. Geneticists, as Hall explains, have become tired of sinister comparisons of their work with *Brave New World* and *Frankenstein,* and deliberately try to set new terms for conceptualizing genetics: 'It's not scientists' intent to reproduce entire human beings, but to exploit bacteria like copying machines' (63). Industrial and frontier imagery serve to revamp the popular imagination of the gene, the geneticist and genetics. The cell-as-copying-machine should set the example for all organisms, including humans, and Hall reverently quotes Robert Swanson, President of Genentech as saying: 'We should all have people working for us who work like micro-organisms. They divide every twenty to thirty minutes and work twenty-four hours a day – just for room and board. What's more, they make the ultimate sacrifice – they die making the product for you'

(277). Scientific goals naturally merge with market goals in the symbolic setting of genetic stories. The gene has developed into the icon of the industrial world; since it is dedicated to the reproduction of human resources *and* capital, it seems to be the closest thing to cloning money.

In contrast to buoyant images of new galaxies and old frontiers, as evoked in *Invisible Frontiers*, the relocation of genetics to industrial compounds has also inspired stories of doom and downfall. A merger of science and money that inevitably leads to corruption has been a recurring theme in popular literature. Robin Cook's fast-paced thriller *Mutation* exemplifies how the new stories of biotechnology may similarly be informed by age-old myths.[51] The book is dedicated to Mary Shelley, who warned us of the 'dangers of creating consciousness without awareness'. *Mutation* recounts the story of Victor Frank, a geneticist and co-owner of a small start-up company called Chimera Inc. He and his wife Marsha, a child psychologist, are unable to conceive more children after the birth of their eldest son. The couple decides to have their genetic material fertilized in vitro, and have the zygote implanted in the womb of a surrogate mother. The in vitro fertilization takes place in the Fertility Center of Chimera Inc., where Victor Frank secretly manipulates the zygote. In order to enhance his child's intelligence, he inserts an 'NGF gene along with a promotor' to cause a mutation in the DNA. Frank does not limit his experiments to his own genetic material, but repeats his experiment in six of his colleagues' in vitro zygotes.

It is not simply the desire for an exceptionally gifted child that prompts the scientist to tinker with genetic material. Commercial pressures to produce fast results and saleable products cause Victor Frank to sell out his academic integrity. As a co-proprietor of a small company that has just gone public, the scientist-turned-businessman experiences the heat of shareholders, who want to see fast returns on their investments. Although genetic experiments are strictly regulated by law, Frank argues that this law does not apply to private institutions like Chimera Inc., and it is this reasoning that allows him to justify his private ventures on moral grounds. The commercialization of biotechnology seems the impersonal malefactor inciting the scientist's act: the lure of profits necessarily leads to calamity and disaster.

The monster resulting from the experiments predictably breaks loose. After the birth of the baby, Marsha immediately perceives something wrong with the child; Victor Junior's eyes look strangely cold – a feature she attributes to the petri dish. At the age of three, VJ (as the child is known) has surpassed his father in intelligence and physical strength, and he appears psychologically mature enough to realize the necessity of hiding his superior powers from his parents and the outside world. Within the next few years, he kills the six children of Chimera employees who could possibly equal or surpass him in intelligence, by inserting in their blood a cancer-triggering gene. Since VJ exhibits an extraordinary interest in scientific research, his father allows him to stroll around the laboratories on the Chimera premises. The child prodigy not only manages to build a lab of his own, with the help of some strong protective friends, but he also invents a gene manipulation process that is far more advanced than the one invented by his father. To finance his scientific enterprise, VJ also creates a secret cocaine lab, whose profits guarantee the pursuit of his scientific endeavour. In a subterranean research laboratory – symbolizing the monstrous underground world of evil and doom – he artificially grows test-tube zygotes who are intellectually inferior but physically strong.

Like *Invisible Frontiers*, *Mutation* expatiates upon the industrial metaphor of genetics; the underground lab is described as a giant factory where large quantities of genetically manipulated embryos are produced and stored. Not coincidentally, this factory is reminiscent of Aldous Huxley's production site where Alphas, Betas and Gammas are gestated in bottles. The underground laboratory, where VJ hides his experiments from the prying eyes of the world, is the immoral counterpart of the 'subcellular realm' in which geneticists descend to enlighten humankind with their new insights. The cause of the ultimate downfall of man is not so much the questionable ambition of a single scientist so much as the fusion between (inherently beneficial) science and (inherently evil) capitalism. A Faustian re-creation, Victor personifies the venture-capitalist who has bartered his soul to the devil and has traded human values for fame, money and power. The marriage between science and money, Cook's popular novel contends, can only result in greed and unscrupulous ambition.

But more than *Brave New World* and *Faust*, *Mutation* redresses the Frankenstein myth. Robin Cook's interpretation of Mary Shelley's plot is rigorously anti-technological, infused with a strong anti-capitalist resentment. It reasserts the uncomplicated equation of science and capital, scientific and capitalist exploitation. Victor Frank impersonates the diabolical nature of technology which, if exacerbated with character flaws like greed and a lust for power, inevitably leads to disaster. The monster merges with its creator to exemplify the inherent malevolence of scientists who pursue scientific glory at the expense of humanity. The reductive reframing of Mary Shelley's story to a scientists-contra-humanist plot invariably informs contemporary Luddite positions on the new genetics. Through its many re-creations, including the influential film versions of the monster story, Frankenstein has itself become a myth and a metaphor,

an image that may concurrently signify the scientist, the monstrous creature, the technology that created it and the society that condones it.[52] The
multivalent plot of *Frankenstein* is continuously reduced in popular
retellings to serve as a warning for what happens if humans tinker with
life, and interfere with their own reproduction. Questions of immorality
have not disappeared from the discussion on genetics, but are redressed to
interrogate the industrial and capitalist dimension of a new technology.

Besides representing an anti-technology stance, Mary Shelley's story
also seems to provide ammunition for an unabashed critique of the male-
centredness of science. When VJ's parents finally discover the underground lab and realize the diabolical potential of their son's work, Victor
Frank is initially tempted to join his son's venture, attracted by the enormous monetary profits that stand to be gained from patents on VJ's gene
manipulation techniques. Yet Marsha convinces her husband that, in order
to avoid total catastrophe and the downfall of humanity, they should blow
up the entire underworld. As in the original Frankenstein story, the creator
and his creation destroy themselves. Only Marsha survives the explosion,
to tell the world of her husband's boundless arrogance; he was so obsessed
with the singular goal of science and profit that he held the higher values
of mankind in contempt: 'Science runs amok when it shakes loose from
the bounds of morality and consequence' (299). The male monster's counterpart is no less a stereotype, although it may be less recognized as one.
Marsha, loving mother and wife of the double monster-creator, personifies
humanity's moral and ethical conscience. She is the typical female who
opens the eyes of the male scientist. Her indictment of human arrogance
sends a warning to ambitious scientists who are so narrow-minded that
they are unable to foresee the long-term social implications of their work.

Frankenstein, as George Levine and others have stated, seems to con-
front us with the 'displacement of God and women from the acts of conception'.[53] Victor Frank's biggest sin is that he defies the procreative
powers solely assigned to God and executed by women. In popular stories
of genetics, this bipolar interpretation – scientists versus God, or scientists
versus women – is frequently invoked to underscore the immorality of the
new genetics. The final removal of reproduction from women's bodies to
the factory or the laboratory, as noted at the beginning of this chapter, is a
recurring topos in feminist warnings of genetics in which Frankenstein
stands for both masculinist technology and industrial politics. Moulded
into the moralist versus scientist frame, the monster prefigures the *Angst*
for extracorporeal reproduction. Reproduction outside the woman's body
taking place in brooders and machines represents the apotheosis of commodification. Just as positive images of space explorations are consistently

complemented by age-old frontier mythology, the symbolic setting of biotechnology in the underworld is coupled with recycled interpretations of popular stories and persistent myths.

THE BINARISMS OF BIOTECHNOLOGY

The performance as well as the scripting, staging and setting of biotechnology appear structured by fallacious binarisms. Sociobiologists launched the image of the 'master' gene in control of its environment – other molecules, cells and organisms. Some feminists picked up the metaphors popularized by sociobiologists and used them to amplify the purported split between genes and bodily environment, men and women, technology and nature, and science and society. The media framed feminists in opposition to scientist-entrepreneurs, hybrid geneticists who took centre-stage under intense public scrutiny. On the one hand they gave rise to sharp indictments – images of noble scientists pacting with the money-devil; on the other hand, they rose to the status of heroes and millionaires, reverently referred to as scientific and financial wizards. Hybrid personae of scientist-entrepreneurs are mostly represented as flat characters. Although the merger between science and commerce rendered these domains inseparable (having been perceived as opposites), we found a recurrent emphasis on the epistemic distinction and hierarchy between science and industry. In popular stories, the wedding between academia and commerce either gave rise to enthusiastic acclaim, or to extensive lamentation of the loss of academic values.

The press played a significant role in the conflation of these two domains. On the level of staging, we noticed how scientists snuggled up to journalists in an attempt to uplift the expectations of a novel industry. The rise of biotechnology stipulated a strategic repositioning of scientists in their attitude towards journalists. As the media provided the most efficient shortcut for scientists in their striving for public recognition, they started actively to court journalists. Taking advantage of the journalist's dependence on news and deadline pressures, biotech researchers increasingly took their research straight to the press, appeasing reporters with colourful images and plastic metaphors. For an industry just 'going public,' with funding heavily dependent on the stock market, public perception became a crucial instrument for their commercial viability. Journalists, for their part, were met by a new assertive group of public relations professionals, who increasingly mediated communication between scientists and the public in general. It seems a contradiction in terms that the 'corporate' image of biotech companies was advertised through the 'academic' integrity of their researchers. Images of products and the persona of the geneticist became vital elements in the race for success and marketability of future products. Professional marketing of products – or concepts of products – required an early mobilization of images.

From the larger, cultural viewpoint, this deep interpenetration of science and industry could be explained as the shift from a modernist culture that reified the emancipatory power of science, to a postmodernist culture that prioritizes commercial values and adopts convenient images to sell this new concept as 'natural'. The notion of 'technology transfer', used to indicate the transformation of knowledge from one domain to another, became obsolete because there is no motion or transfer in a harmonized public/commercial realm. As Fredric Jameson observed, the emergence of postmodernism is closely related to a consumer culture in which the signifier replicates the logic of late capitalism.[54] Translated to the genetics industry, this cultural shift is noticeable in the coalescence of idea, image, (projected) product and production process. The idea of the cell as a microbal factory feeds the scientific research process which is in turn nurtured by a commercial incentive to produce large quantities of marketable products. Images are as important as products, because they provide the basis on which capital is raised. The image, in fact, is capitalization: pivotal to the commercial success of biotechnology is the concurrent shaping of research, marketing and public relations agendas.

Yet in spite of the apparent postmodern fusion between science and commerce, the outward, symbolic division between the two realms is a *sine qua non* for genetics' trustworthy popular image. The image of neutral, objective science formed a unique selling-point in the marketing of biotech products. The ostensible split, albeit cosmetic, between disinterested science and for-profit industry is repeatedly foregrounded in advertisements and public relations material. Scientists and journalists, while insisting on their distinct professional responsibilities and activities, are mutually involved in each other's designated roles as image producers. The hybrid physical setting of the biotech industry – the emergence of science parks close to universities – is symbolically offset by uniform spatial images of galaxies, frontiers or classic underworlds of doom and downfall. And the new industrial images of genes, geneticists and genetics are consistently realigned with old myths and recycled imaginations. Those conventional images and imaginations keep intact the very binary categories and hierarchical oppositions that inform the so-called postmodern border crossings. In the next chapter, we will see how the diffraction of scientific and commercial activities continues in the creation of a new medical image for genetics, while conventional images reinscribe traditional binarisms into the popular imagination.

5 Biophoria: The Human Genome Project

MAPPING AND READING THE GENOME

In 1985, Chancellor Robert Sinsheimer of the University of California, Santa Cruz (UCSC), convened a meeting of renowned biologists to discuss the possibilities of what later turned out to be an audacious but practicable plan: the sequencing and 'mapping' of the entire human genome. In a position paper published by the UCSC Biology Board, the Santa Cruz scientists stated that although it is not at all clear that 'knowledge of the nucleotide sequence of the human genome will, initially, provide deep insights into the physical nature of man', it did not seem any more far-fetched than a moon landing.[1] After Sinsheimer had ignited the flame, Walter Gilbert and James Watson took over the torch and started to campaign for political and public support. The lobby to convince Congress and tax payers of the merits of a total investment of three billion dollars until the year 2005 took less than five years; the Human Genome Project (HGP) was approved by Congress in 1990.[2] The HGP indeed became the biological version of the 'man-on-the-moon' project, as it pulled together a lot of the scientific community.

Though the HGP started as a technological vision, molecular biology's promotion into the league of Big Science could materialize only through a coalition between science, industry and government agencies. A concerted effort to inventory the human genome, however, encompassed much more than the execution of a plan; it entailed the development, distribution and implementation of a way of thinking about human life. Assuming that a disposition towards a large number of diseases is stored in the genes, the HGP proposes systematically to apply the tools of molecular biology to the entire complement of DNA in a human cell, on the premise that once we know the complete genetic make-up of the human body, we can diagnose and predict congenital aberrations. A full inventory of the correct order of nucleotides on the human chromosomes, in this line of reasoning, is a precondition for prevention and cure of all genetic predispositions. The Project's aims are multiple and various, directed at theoretical as well as at practical research. In the public mind, however, the HGP stands out as a scientific enterprise with a primarily *medical* goal: to find the 'fixed' pattern of our genotypical composition in order to 'fix' congenital diseases. Genetic screening had prepared the public mind-set for the next great step in research: gene 'therapy' promised to be the pinnacle of curative medicine. The 'geneticization' of society seems to be the flip side of the 'medicalization' of genetics. Despite its abstract theoretical goal, human genome mapping managed to generate overwhelming enthusiasm among the general public. A true 'biophoria' in popular representations of the genome project resulted from a strong injection of powerful images and imaginations in the public domain as well as from a changed dynamic between the various professional groups involved in the staging of this scene. Needless to say, the Project also invoked disapproval and criticism, to which I will pay extensive attention in the next chapter.

As the Human Genome Project gradually gathered public and political support, we can see a decisive change in the popular images that helped disseminate genetic knowledge. Images of the gene-as-factory and manager-controller, which had accompanied the birth of biotechnology, rapidly lost ground in the 1990s. The gene metamorphosed into the 'genome', genetics into 'genomics', and the geneticist became an amalgam of a molecular biologist and a computer scientist. More intricate mental concepts were needed to imagine the genome as a digital inscription of the body's genetic make-up. It is not a coincidence, for many reasons, that the new genetics derived its primary images from the burgeoning computer industry, as the development of genomics was thoroughly dependent on the emergence of computers. The computer, despite its relative recent introduction as a biological research tool, was not a completely novel image in the language of genetics. Since the 1960s, the metaphors 'information' and 'code' had provided an explanation for the unidirectional flow of 'messages' from RNA to protein. Yet in the course of the 1980s, a paradigm shift in molecular biology opened up the cellular message to two-way traffic. As Evelyn Fox Keller has explained, the view of DNA as loci of information, and RNA as its 'messengers' or transmitters, was replaced by more complex interactional models: 'Notions of networks and organizational complexity, borrowed from cyberscience by theoretically minded biologists were reimported to biology in an effort to revitalize older organicist conceptions of development.'[3] Cyberscience – a compilation of information theory, cybernetics, systems theory and computer science – provided the new conceptual frame for genetics, and enabled scientists to talk about human DNA as so-called data in an ingenious network of organic units. Whereas 'information', in the 1960s, had served as a metaphor, it now became material inscription. The notion of a genetic 'program' entails a paradoxical loop:

the program requires the very proteins that it generates in order to go on generating them. Circulation, rather than a linear flow of information, provided the vector for the dissemination of meaning. Now that the computer dominated the genetic imagination, the body became part of an informational network.

The computerization of genetics triggered a whole new set of related images which, on the face of it, seemed reinvigorated or updated by computer language, but which turned out to be far from innovative. First, the reconceptualization of the genome resulted in the distribution of an image crucial to the materialization of the HGP: the creation of a 'map'. The effort to determine the sequence of all three billion A, G, T and C base pairs that comprise the human genome, in fact consists of two projects: genetic mapping and DNA-sequencing. Genetic maps show the relative positions of thousands of genes on the chromosomes – the repositories of genetic information. Gene mapping is further detailed by DNA-sequencing, which determines the exact order of base pairs along the chromosome. Combined, genetic linkage maps and physical sequential maps 'chart' the human genome on a miniature scale. The axiomatic implication of the metaphor is that the 'map' enables the localization of specific genetic aberrations or diseases. Ingrained in this configuration of the genome is the idea that disease can be 'localized' as a defect or an irregularity in the gene sequence. A 1986 article in *Science*, discussing the proposal to sequence the entire human genome, forecasts the possibility of pinning down the 'location of, say, the cystic fibrosis gene to one particular 40 kb fragment'.[4] The ultimate map is seen not just as a metaphor but as an *actual guide* to the layout of our DNA make-up, so that deviances or defects can be easily pinpointed.

In both scientific and popular discourse, the term 'mapping' frequently surfaces to elucidate the activities and goals of genomics researchers. In fact, the term is used so often that its figurative meaning escapes the reader's notice and reaches the stage of demetaphorization. Science journal articles and newsletters are replete with map-related jargon: geneticists talk about 'sequencing a region', 'guideposts', 'disease loci' and 'genetic markers' as if there are actual sites to which illnesses can be reduced. The very concept of 'disease loci' noticeably affects the techniques used by scientists in the mapping process. In 1990, *Human Genome News*, the official newsletter of the HGP, announced two new techniques to speed up the location of genetic markers.[5] 'Sequence tagging sites' and 'reverse genetics' promote detection and examination of markers that cause defects, and subsequently identify the flaw in a string of codings of human DNA. From 1990 onwards, discoveries of 'flawed gene locations' dominate the front pages of the newsletter. The gene which locates neurofibromatosis is reported in September 1990, the gene for cystic fibrosis in January 1992, and the gene that is held responsible for myotonic dystrophy in March 1992. Invention and stimulation of methods like reverse genetics tacitly change the primary goal of the mapping project from defining the ideal (healthy) human being into defining its diseases or flaws. These goals might seem like two sides of the same coin, yet a strong emphasis on determining aberrations in the genetic make-up yields a view of the body as the flawed version of the perfect code, and concurrently holds the promise of an easy genetic fix.[6]

In national and local newspapers, discoveries of gene locations are hailed as successes, even though finding the location of a flawed gene does not imply the slightest chance of a cure. Still, the image of the genome map implicitly ratifies the need for genetic tests and genetic therapy. In popular gene stories, the projected map will provide the model for explaining health and disease. As *Time* magazine has predicted, by the year 2005 the wall of every doctor's office will prominently display a colourful chart of the human genome, hanging next to the physician's graphical illustrations of skeletal parts and organs:

> It will depict twenty-three pairs of human chromosomes and pinpoint on each one the location of genes that can predispose people to serious diseases. By then, scientists will have sequenced the three billion chemical code letters in that strand, giving them the ability to read nature's complete blueprint for creating a human being.[7]

Like a geographical map, a genome map will help physicians to navigate the wild terrain of human DNA; without such aid doctors would be lost in the dark.

The way in which the map metaphor is appropriated by popular discourse reveals how it instills particular *idées fixes* in the public mind. A 'map' is programmatic in a sense that it figures the human body as a predetermined route, a methodical arrangement of DNA-fragments. It also suggests that diseases and defects are in essence retraceable to genetic predisposition. And, not unimportantly, the map metaphor suggests that it is only a small step from charting the human genome and locating genetic aberrations to correcting those diseases. Or, as *Time* magazine explains: 'Locating a gene from scratch is like trying to find a burned out light bulb in a house located somewhere between the East and West Coasts, without knowing the state, much less the town or street the house is on.'[8] Locating a 'burned out light bulb' implies that a genetic defect can be easily fixed by screwing in a new one. Although map images do not emanate from an intentional collocation of semaphores, they easily lend themselves to the dissemination of deterministic views.

A second metaphor that makes a remarkable comeback in the computerized context of genomics is the image of the genome as an 'alphabet' or 'language'. Parallels between cellular systems and the alphabet first became popular in the 1960s, when molecular biologists used the metaphor as an heuristic device to expound complex theories about the working of DNA. At that time, linguistics provided symbolic leverage to popularize the 'grammar of biology' for a general audience. Three decades later, the notion of 'coding' refers to sequences of digital data. What started as a comparison between DNA-molecules as the four basic letters of the human alphabet now resurfaces as a demetaphorized way of thinking about complex cellular systems in terms of information processing and digital decoding. Along with this semantic change, the disciplinary context for applying the language metaphor shifts from linguistics to informatics.

As computerized information systems quickly gained ground, this new conceptual framework allowed for notions of complexity and circularity. However, the images that evolve along with information processing systems reflect anything but complexity. While the human genome is increasingly *figured as* a 'blueprint', it has *de facto* become a collection of digital units, stored away efficiently on a compact disk. Walter Gilbert, one of the co-founders of the HGP, proposed the idea that the complete sequenced human genome, inscribed on a compact disk, will contain all our essential information: 'Three billion bases of sequence can be put on a single compact disk, and one will be able to pull a CD out of one's pocket and say "Here is a human being, it's me".'[9] In Gilbert's evocative and appealing image of the compact disk containing an inscription of a person's genetic make-up, we can notice an unproblematic collapse of digital, organic and metaphysical signifieds of the body. 'Me' refers simultaneously to digital units as a representation of the genetic body, an electronic inscription and an immaterial 'essence' – an identity or soul. Gilbert's metaphor of the compact disk does more than 'update' the familiar equation of 'life' with molecular structure, propelled by Watson and Crick. The parallel between 'life' and digits materializes in a palpable, digital artefact – almost a household product.

Digital inscription devices and storage systems are not merely technological 'upgrades' of old information and language metaphors. The concept of the material body develops in conjunction with inscription, technology and ideology. Sequencing genes – the activity that constitutes the core of the HGP – now fully consists of processing digital information. The idea that the human body can be coded in a decipherable sequence of four letters, and hence in a finite collection of information, is based on the epistemological view that computer language – like molecular 'language' – is an unambiguous representation of physical reality. Whereas the metaphor of mapping suggests an analog representation – a linear registration of a flat surface – the sequencing of genes ushers us definitely into the digital era. Digital encoding differs from analog recording in imposing a language of zeroes and ones, combined into great complexities, onto the human material body.[10] This digital 'registration', however, is not merely a representation of molecular language, but more like a translation into a different dimension. Through the inscription of 'DNA-language' in digital data, the body is turned into a sequence of bits and bytes whose function is no longer exclusively representational.

Besides being a representation of an organic body, the digital genome data now also function as the material inscription of a 'model' body. Digital data can be recorded and rearranged to form an 'ideal sequence', the gold standard which has no referent in reality. A 1990 issue of *Human Genome News* addressed a question often asked to scientists: Whose genome is being used?[11] The ultimate genome map, as the article explains, will be a sort of composite. People are ninety-nine per cent similar in their genetic make-up, and 'because these small differences vary from person to person, it does not matter whose genome it is'. The human genome map is the map of a hypothetical person, a standardized model which corresponds to no one in particular, but is derived from the tissue of thousands of donor bodies. In other words, the genetic blueprint is not a transcription of a real body, but the projection of an 'ideal body', constructed as a univocal formula, a mathematical language. Biotechnology combined with informatics transforms the body as an (organic) object of knowledge into an ordered collection of biotic components – itself an 'image' or 'concept'. Several critics have argued that the body ceases to exist as an object or entity because the digitized sequencing of DNA turns it into an inanimate compound of encoded numbers.[12] Or, as Keller aptly sums up:

> As a consequence of the technological and conceptual transformations we have witnessed in the last three decades, the body itself has been irrevocably transformed, perhaps especially in biological discourse. ... The body of modern biology, like the DNA molecule – and also like the modern corporate and political body – has become just another part of an informational network, now machine, now message, always ready for exchange, each for the other.[13]

Although genetics and information technology have fully merged to construct and represent complex organic systems, the notion of 'reading' seems naturally to imply automatic decoding – a task performed by a CD-player rather than a human reader. Such narrow definition of reading prohibits any interpretive activity, and denies the presence of inherent ambiguity in DNA-strings of information. Yet the same 'words' or DNA-sequences may have different meanings in different contexts, and multiple functions in a given context. Besides, the concept of interactivity suggests that the actual value of 'knowing' the composition of these digital strings is for gene mappers to be able to manipulate them infinitely. Interactivity also connotes human power over a man-made and man-controlled digital system; the words 'computer program' perfectly synthesize the notions of 'programmed' and 'programming'. The upgraded metaphors suggest that digital information can exist only in relation to a genetic referent. However, the nature and structure of computer language is such that it may create relationships among pieces of information outside the confined borders of genome mapping. Genetic data, in digitized form, are convertible to other informational contexts. Someone's sequenced DNA, for instance, may be of value to insurance companies or mortgage bankers. Digital information never consists solely as a representational system, but can function as an autonomous signifier.

The hybridity of gene blueprints as concurrently material and representational inscriptions lies at the heart of genomics practices and policies. New digital metaphors reflect this dual nature, as they allow strategic mixing of representational and ontological meanings. Terms like 'editing', 'copying', 'deleting' or 'retyping' are used in their denotative meaning: genetics is no longer explained *in terms of* (computer) language, but is fully integrated into informatics. Though the merger may appear merely transient, its consequences are not. When in the late 1980s a debate erupted over the question whether sequences of encoded DNA-material could be patented – and thus owned – by scientists or corporations, the images of genomics offered a new venue into thinking about patents. Since strings of coded DNA are merely an imprint of 'natural' phenomena, no one can obviously claim ownership of identified sequences. Nevertheless, scientists and corporations can own the instruments used to 'edify' nature.[14] A lawyer explained in *Human Genome News* that scientists can exclusively claim the lucrative fixes to which this knowledge presumably gives access.[15] To achieve this aim, we do not need to change any laws, we only need to change the way in which we conceptualize sequenced strings of DNA. Geneticists should no longer think in terms of patents but of 'intellectual property rights', the lawyer argued. Whereas patents refer to actual products, the ownership of sequenced DNA-material is legally provided for as copyright. Obviously, the goal of the HGP is not to make the body 'legible' but to decipher its script so that editing comes within the limits of scientists' scripting abilities – an image reminiscent of the ancient idea that man should take evolution into his own hands. The compact disk analogy, both as metaphor and material practice, does not just change the popular image of genetics; it actually modifies a way of thinking about the human body and the way genetic knowledge and information can be stored, owned and accessed.

Paradoxically, while usage of the terms inscription and information in the context of genomics has been demetaphorized, traditional interpretations of these images appear tenacious. When talking about 'reading' or 'mapping' the human genome, scientists still hint at unidirectional deciphering, adhering to the notion that all essential information is stored in the DNA-molecule. Indeed, computer-related terms allow for expansive interpretations to cover complexity and interaction between molecules, but their popular versions fail to exemplify new insights into the circularity of genetic information. So the 'fixed' notion of a body determined by its genetic predisposition remains intact, while genomics enables 'fixability' of the genetic pattern by 'mapping' the digital code. Demetaphorization, or the materialization of a formerly representational system, seems to evolve along with a reinforcement of the old metaphorically induced signifieds. The persistent images of genome reading and mapping as activities that involve encoding and decoding of information are amplified through the plots and stories of genetics invoked to explain the need for 'genome mapping'.

HUMAN GENOME PLOTS

In popular versions of the HGP, the 'map' and the 'blueprint' are paired off consistently with a number of historical, cultural, mythical or religious narratives. Geographical mapping, in the digital era, obviously involves the use of sophisticated computer technology and satellites, but genome 'mapping' is rarely associated with high-tech equipment. Instead, scientists and journalists often revert to fictional or historical expeditions, in which heroes set out to discover and map new lands. Non-fiction books on the HGP are replete with references to 'treasure hunts', 'pioneer adventures' and 'jungle expeditions', stories that rack up the excitement of discovery.[16] The human genome map is likened to a 'treasure map' which will lead you to the place where the treasure is hidden, even though nobody still knows what that secret really is. The 'gene map' is also compared to the search for unknown territory, as evidenced by a description in

Time: 'Like cartographers mapping the ancient world, scientists have been laboriously charting human DNA, yet long segments of the genome, like the vast uncharted regions of early maps, remain terra incognita.'[17] All these adventure plots generally connote discovery and excitement, but they are not very specific in their attachments of ideological values to the Human Genome Project.

As we saw in the previous chapter, biotechnology adventures were frequently wrapped up as conventional frontier stories. The Human Genome Project is related to specific historical 'map making expeditions'. The act of recording newly found lands on paper signifies a derivative activity, providing symbolic proof of the expedition's success in broadening horizons and conquering expansive living space. Lois Wingerson, in her popular non-fiction account, compares the HGP to Columbus's 'discovery' of America:

> Like Columbus leaving the coast of Spain, the explorers of the human genome cannot yet see behind the horizon. ... The rest of us are like Ferdinand and Isabella. We are paying for the voyage, and we know it's under way. But we are left behind on terra firma. All we can do now is wait to see what the explorers bring back to us.[18]

An analogy to Columbus's voyage helps Wingerson explain the unknown in terms of historical triumph – journeys which have brought America prosperity and glory. According to *Business Week*, the work of gene mappers 'rival[s] that of Balboa, who crested a mountain range in Panama to see – and claim for Spain – a whole new world'.[19]

A second historical event frequently invoked to justify the need for an abstract mapping project is the penetration of the American West. Gene mappers are referred to as the 'Lewises and Clarks who wanted to reconnoiter the genetic continent and locate valleys, lakes, rivers and fields (i.e. genes) that could be exploited'.[20] In popular discourse, genomics seems to revive old colonialist notions of discovery and conquest. The 'discoveries' of America or the American West are presented not as an interpretation of history, but as uncontroversial historical facts. By means of analogy, the basic benevolence and economic pay-off of neo-colonialist expeditions is projected onto the 'genetic frontier'. The neo-colonialist myth is remarkably similar to the old, as the need to discover and own dark continents is presented as a 'natural' impulse.

The genetic frontier, however, is different from most historical explorations of new territories. While the expeditions of Columbus and Lewis and Clark were aimed at discovering and appropriating foreign lands, the genome map does not offer new land or inhabitable space; it only offers a new way of thinking about the human body. This way of thinking is summarized in a frequently quoted phrase from James Watson: 'We used to think our fate was in our stars. Now we know, in large measure, our fate is in our genes.'[21] And Walter Gilbert formulates its goal poignantly as: 'To recognize that we are determined, in a certain sense, by a finite collection of information that is knowable will change our view of ourselves. It is the *closing of an intellectual frontier*, with which we'll have to come to terms.'[22] The equation of inhabitable space and knowledge, both figured as land waiting to be conquered, reveals the essentialist underpinnings of this venture. According to Watson and Gilbert, the Human Genome Project is the search for ultimate knowledge of the human body and the self, the key to predicting our personal futures which are laid down in the stars, the cards or the maps.

Since the terrain to be conquered in genomics is intellectual rather than physical, scientists and journalists also revert to mythical plots. Walter Gilbert's essay is significantly titled 'A Vision of the Grail' as he compares the HGP with the quest for the Holy Grail. Yet the moral of the original Holy Grail story is incommensurate with its popular reframing. The plot of this myth – twelve Knights sent out to retrieve a mysterious treasure that exists only in the various imaginations of its hunters – ends with the recognition that the Holy Grail is in fact an unknowable essence, symbolizing the ultimate immateriality of human knowledge and happiness. By contrast, the goal of the Genome Project is to inscribe the essence of human life, and hence to render it material. Despite this incongruence, Gilbert's hyperbolic statement on the human genome as the Grail resonated widely through popular discourses.

Similar to the popular stories accompanying the map metaphor, the image of 'reading' has given rise to a revival of cultural images and narratives. From the new digital context of the genome as a 'computer program' you would expect a digital upgrading of metaphors relating to 'information' and 'networks'. Instead, information pops up consistently in relation to conventional notions of language and reading. Scientists and journalists wield a palette of culturally imbued terms, ranging from the good old 'alphabet' and 'translation', to the newer 'encyclopedias' and 'libraries'. A common contention expressed in popular magazines is that, as soon as scientists have sequenced the three billion chemical code letters, 'they will have the ability to read nature's complete blueprint for creating a human being'.[23] The genetic composition of the body can be spelled out in four single letters, as *U.S. News and World Report* explains: 'Just as the English language uses twenty-six letters in various combinations to make thousands of different words, different arrangements and repetitions of these four bases make all of the body's genes.'[24] More than once, the effort to collect the human genome is likened to the assemblage of a huge encyclopedia, such as the classic *Encyclopaedia Britannica*, or a giant dictionary containing all the world's idioms or a huge reference work: 'If the human genome is an encyclopedia divided into twenty-three chapters (chromosome pairs) each gene sentence is composed of three-letter words which are in turn spelled by four molecular letters called nucleotides.'[25] Parts of the genetic code that have been sequenced are stored in so-called 'gene libraries', a term that frequently surfaces not only in popular magazines but also in scientific journals, without the quotes to denote their figurative meaning. As in a library, some books or genes will rarely be consulted, because 'no one uses all of it, but each part is used by someone at some time'.[26]

Cultural images purportedly serve to elucidate the mechanisms of nature. At the same time, however, they reinscribe concepts of nature as well as of culture. 'Reading' in relation to DNA-sequences means decoding encoded information; it infers there is a natural order of things, that genes occupy a fixed place in the body, and that they are merely there waiting to be deciphered. Once the laws of nature are recorded in the 'reference guide', one can endlessly retrieve and copy stored information. Moreover, the explanation of the HGP as the construction of an encyclopedia and a library impart the notion of genome research as a humanist undertaking, a monumental effort to build up a 'body of knowledge'. Comparison with the Bibliothèque Nationale in Paris suggests the tremendous cultural value attached to the word 'collection'; both the HGP and the national library strive to collect, sort out and store valuable information, guarding our scientific and cultural heritage along with our genetic heritage. Almost imperceptibly, the terms 'information' and 'knowledge' get mixed up in the use of these plots. Just as storing information is not the equivalent of accumulating knowledge, collecting DNA-sequences does not automatically result in a better understanding of the 'essence of human life'. Modelling nature after culture is the strategic equivalent of Dawkins's attempt to model culture after nature; both betray a view of culture as a massive, static body of knowledge that, once accumulated and systematically stored, can be endlessly retrieved.

Besides historical, mythical and cultural narratives, popular stories of the HGP are structured by religious plots. Building on the root metaphor of reading nature as a book, we find frequent references to the HGP as the production of the 'Book of Man' or the 'Book of Life'. Underlying this notion is the idea that life can be seen as 'written' somewhere, and the true mission of humans is to find the 'true' meaning of that script. Walter Gilbert, who had declared the HGP the 'holy grail' of genetics, also noted at one point that the sequenced genome is the 'ultimate answer to the commandment, "Know thyself".'[27] The HGP is often referred to as the 'code of codes', implying that the univocal pattern emerging from this giant enterprise will offer the key to understanding human illness. This code is at once secret and transparent; it needs to be deciphered before it can be read. And reading, in line with dogmatic religion, entails the obedient act of deciphering rather than a complex signifying practice. The stories that science produces are obviously bound up with theological narratives that are deeply ingrained in our cultural practice.

Through the attribution of 'fixed' plots, the Human Genome Project is promoted as a way to ferret out nature's secret, or, as Keller calls it, a 'rendering of what was previously invisible, visible – visible to the mind's eye, if not to the physical eye'.[28] Biology has become more and more a science that does not tolerate secrets, and, in the retelling of its own success stories, 'the focus inevitably shifts from the accomplishments of molecular biology to the representation of those accomplishments' (42). Reflected in the metaphor of the map and the blueprint, we can sense a strong inclination towards closure, towards unravelling the secret of life – as Watson had boasted in 1953 – and towards a desire to penetrate the realm of the invisible to turn it into the knowable. The reappropriation of mythical quests amplifies the image of genome research as a holy vocation, and reinforces the idea that the essence of human life, the composition of the body, is an object that can be and should be thoroughly demystified.

The strategic recovery of the mapping and reading metaphor by far exceeds the goal of exemplification proper. Configuring the human genome as a map or blueprint appears pivotal in the construction of a new kind of genetic thinking. Disease can be retraced to an actual genetic locus and can be decoded once the key has been found. The promise of genetic 'therapy' – the pinnacle of heroic reparative medicine – is prefigured in genome metaphors. A new and abstract domain of knowledge is discovered and conquered in the name of old and historically beaten tracks. Mythic figures and historical voyages – the Knights Templar, Columbus, Balboa – help visualize the abstract mission of geneticists, turning their invisible theories into palatable goals. Given the genome's endless possibilities for reappropration and reinterpretation, it is surprising to see such massive restoration of ancient images and plots.

THE GENETICIST AS MEDICAL HERO

A similar incommensurability between 'upgraded' images and conventional metaphors and plots proliferates in popular configurations of the geneticist. The concerted effort to sequence and store all of the expected 3.5 billion pairs of DNA requires close co-operation of molecular biologists and computer scientists, who are capable of operating the hardware and software involved in processing huge databases. Specially designed computers enable scientists to encode large quantities of information at great speed. In popular accounts on the HGP, remarkably little attention is paid to the computer hardware and software that is so crucial to the Project. Within the HGP, a special 'Joint Information Task Force' co-ordinates the development and implementation of new computer programs and electronic sequencing strategies. In early issues of *Human Genome News*, we still find reports on how 'computer scientists join molecular biologists in workshops'. In later issues, reports on the information taskforce become scarce, and their activities seem to be subordinated to the 'real' genome sequencing. Yet a closer reading reveals that the development of hardware has not taken a back seat, but that informatics has fully merged with molecular biology, creating new disciplinary fields in the process. Labels like 'bioinformatics' and 'genomics' refer to the joint efforts of molecular biologists, engineers and computer scientists, without distinguishing between their respective disciplines. New professional categories and hybrid disciplines emerge at the crossroads of various sciences and technologies, which appear to occupy strategic positions in the development of marketable knowledge and products. 'Imagenetics' or chromosome painting is a fluorescent staining technology that combines the work of medical scientists and computer technicians.[29] And 'database curation' introduces a new type of professional ('similar to a museum curator') whose job it is to decide which information should be stored and how.[30]

In popular representations, genome mappers are rarely depicted as information specialists. A far cry from the wizardly lab worker or the savvy scientist-entrepreneurs, the prevalent image of the geneticist involved in the Human Genome Project or gene therapy is that of a doctor in a white coat. Most popular stories focus exclusively on the medical applications of genome mapping, and ignore the multiplicity of technical skills required for its execution. An extensive profile in *Time* (1994) illustrates the homogeneous depiction of genome mappers and gene engineers as medical heroes.[31] Francis Collins, appointed in 1994 to head the Human Genome Project, and French W. Anderson, director of the Gene Therapy

Laboratories and leader of the first gene transfer experiment, are presented as the designated pathfinders of the new medical revolution. Anderson and Collins, besides being scientific superstars, are also endowed with divine qualities, as their primary mission is to save the lives of innocent victims of congenital disease. Full-colour pictures of Collins on a motorbicycle and Anderson exercising Tae Kwon Do, a Korean martial art, take up almost half the space of the profiles, foregrounding their gung-ho, fighting spirits. Francis Collins is not only described as a 'relentless hunter of disease genes' but also as a 'devout Christian' whose selflessness rivals that of a saint. The *Time* reporter relates how Collins once travelled to Nigeria to treat patients in a small missionary hospital:

> Once in Nigeria, he agonized over whether his presence there made any difference at all. Then a farmer appeared, suffering signs of imminent heart failure. Collins dared a procedure he had never tried before, plunging a needle deep into the man's chest to draw off the fluid that was apparently pressing on the heart. 'Dr. Collins,' the patient said later after recovering, 'I know you're wondering why you're here. I believe you were sent here just for me, because without you I would have died' (49).

Obviously, Collins's missionary zeal and Christian devotion have nothing to do with genetic engineering. This scene, with its biblical overtones, sets the stage for the image of the geneticist as a crusading curer. Collins believes in gene therapy and is ready to lead the genetic crusade, joined by many people who 'are believers with him', according to *Time* magazine.

Extensive profiles of scientists are not all that common in general interest magazines, and the ones of Anderson and Collins are striking in sheer quantity of coverage. Few Nobel Prize winners ever receive the attention accorded to these two geneticists, increasing the prominence of genetics as a whole. Genetic engineering, in the form of these profiles, gets a human face, as the scientists' personal qualities are projected onto the entire scientific field. Anderson's dedication to curing patients is stressed more than once; his 'human compassion for our fellow man who needs help now' seems his only incentive to push gene transfer experiments. Scientists who have begun to think about ways to repair the molecular system where it shows serious defects in its design, are called 'copy editors' or 'gene correctors'. Their work represents the pinnacle of curative medicine: they attack illness at its roots, hence preventing physical deterioration. The very notion of 'DNA as text' allows the idea of altering genetic texture as a way of correcting nature's mistakes.

But at least as important as the images of geneticists are the images they produce. New hybrid professions employing advanced visualizing technologies yield enlarged microscopic pictures of cells and genes that not only serve as illustrations of gene mapping technologies, but increasingly as eye-catchers for popular magazines. A *Time* cover story on the genetic basis of cancer and the ensuing promise of better therapies juxtaposes photographs of gigantic cells, glowing telomeres on chromosomes and multiple tumour suppressor genes with pictures of the scientists who produced them.[32] The cover itself consists of a double-page full-colour picture of a breast-cancer cell, whose chemicals dissolve surrounding tissue. Photographic enlargements of cancer cells and genes appear aesthetically pleasing complements to pictures of smiling scientists. Illustrations of cancer cells literally 'blow up' the threat of congenital disease to gigantic proportions, and the small pictures of scientists next to enormous cancer cells insinuate that they are battling an enemy of enormous proportions. Many popular stories also feature illustrations of digitally produced DNA-samples, or a computer printout from a gene reading machine displaying its DNA-codes. The digital-genetic image as an autonomous 'quasi-object' comes to serve as visual proof of a self-effacing reality in the theatre of representation.[33]

THE TRIPLE HELIX

The dominating presence of the medical geneticist in the Human Genome Project does not mean that the scientist-entrepreneur prevailing in the biotechnology business has completely disappeared from the scene. At the onset of the biotechnology boom, the university-industrial complex had become an important ideological stake in the battle between various professional groups to define the meaning of the new genetics. The gradual implementation of the HGP added a third contender to the staged script: government became a major player in the genomics game. Relations between universities, industry and government agencies – also referred to as the 'triple helix' – shape the research agendas of genomics, determine networks of communication, and define in what (virtual) settings cooperative research activities are performed. A co-evolution between new technological developments and their cognitive and institutional environment definitely changed the knowledge infrastructure, and this change in turn refined the dynamics between professional groups involved in the process of image-making.

The image of the 'network' not only functions on a conceptual level, but extends into the institutional-political context in which the HGP materializes. Underlying the organization of the HGP is a web of individual

scientists and research institutions who became aware of the surplus value
of their combined efforts. In the non-fiction annals of the HGP, its genesis
is explained as the realization that numerous scientists working in various
locations were all working towards a communal goal: understanding
human disease. The HGP is thus not a new research project, but a fusion
of existing projects. Laboratories in dispersed places such as Berkeley,
Los Alamos and Washington DC were already trying to find genetic patterns that could account for diseases like cystic fibrosis or Alzheimer's. In
1990, the National Center for Human Genome Research (NCHGR) was
founded to co-ordinate and attune research activities in a number of laboratories across the USA. Both the Department of Energy and the National
Institutes of Health (NIH) got involved in the organization of the HGP; the
Department of Energy, like the National Laboratories, seemed eager to
share in the euphoria of a big science project, rather than putting their bets
on dwindling defence budgets. It is remarkable how quickly the idea of
pushing biology into the 'big science' league built up momentum, and
how smoothly unwieldy bureaucracies accommodated the implementation
of this giant project. In his detailed account of how the Human Genome
Project gradually gained political clout, Robert Cook-Deegan relates how
more and more bureaucracies got involved in the process.[34] After the
National Academy of Sciences (NAS) approved the Project and issued
budget recommendations, the NIH managed the first HGP budget, taking
over the lead from the Department of Energy. Cook-Deegan summarizes
the initial inter-agency squabbling and the emerging bureaucratic organization as follows: 'The HGP knitted together a disparate and widely dispersed genetics community, weaving an informal network of electronic
and personal ties' (122). The network image equally applies to the collective efforts of researchers and institutions involved in the Project as to its
projected outcome.

Whereas the HGP was initially funded by tax payers' money, from 1992
onwards private industries got involved in the effort to map the human
genome. An editorial in *Science* records that the Project, almost overnight,
evolved from solely a public effort to a joint public and private undertaking: 'More than thirty leading genome scientists are cutting deals with
venture capitalists and new companies are springing up around the
country. And this time around virtually everybody is welcoming the
trend'.[35] A highly publicized exponent of this trend was the joint venture,
in 1994, between the Institute of Genomic Research, headed by Craig
Venter, and Human Genome Sciences, a for-profit company directed by
William A. Haseltine. They arranged a contract with pharmaceutical giant
Smithkline Beecham, who received the exclusive rights to market any

products emanating from 'copyrighted' gene sequences. Sequenced data were thus no longer public property and, by 1995, the involvement of private companies in basic research seemed only natural.[36] The unifying goal – finding the scientific key to curing a score of congenital diseases – apparently overrides all previous local and 'special' interests. Economic stakes are conspicuously absent in the excitement of a unified mission for universal health. Historical nationalistic claims to ownership of newly discovered territories are aligned with commercial claims to 'copyrighted' DNA sequences.

The merger between universities, industries and government agencies evolved as the HGP took shape. In the early years of biotechnology, bureaucratic committees and government agencies were seen as a major impediment to the expansion of the university-industrial complex. Yet as the HGP developed from an audacious plan into a ramified network of co-operating research groups, 'government' became less an obstacle to defeat, and more a research partner. In the many studies that outline the gradual coalescence of the distinct interests and goals of universities, industry and government, a fourth factor in the shaping of the 'triple helix' is often underexposed. The media, as we have already seen, played a significant role in the merging process, but in line with this institutional obfuscation, the media itself became increasingly more porous as a place for public discourse. From the framed stories of popular biophoria, professional distinctions between journalists, public relations managers and scientists are far from clear.

A joint goverment and industry involvement in the realization of the HGP can not be explained without the concomitant cultivation of public support. From the very beginning, the promise of gene therapy comprised an important rationale for mapping the human genome. In the past decades, the prospect of replacing stem cells in a living human body had elicited strong disapproval, but in the wake of biophoria, the very idea of gene manipulation quickly gained ground. A poll conducted in 1992 showed an almost complete turnaround in public support for experiments with gene transfer since the early 1980s.[37] The metamorphosis in public opinion seems at least partly due to the medicalization of genetics; 'genetic therapy' became the preferred semantic coating for what was previously known as 'genetic engineering' or 'manipulation'. The semantic transformation of genetics into extracorporeal computer-based genomics undoubtedly assuaged public opinion. When we look at journalistic embellishments of genome euphoria, we cannot fail to notice a complete incorporation of information processing into the language of genetics. 'Gene therapy' is commonly described as an attempt to find a healthy copy of a missing gene and insert it into the erroneous cell. Now that doctors had joined scientists in their efforts to ban congenital disease, the prospects of transgenic people appeared less daunting than when previously worked on by mysterious geneticists in their DNA labs. Although technical barriers had prevented gene therapy from happening at an earlier stage, the progress in this field pressured scientists and medical doctors to knock on the doors of federal legislators. Experiments in mice and cattle had proved the treatment to work, and Food and Drug Administration (FDA) officials were urged to change regulations prohibiting experiments with gene transfer in human beings.

Obviously, the implementation of gene 'therapy' required more than technical skills: it also required massaging of public officials and public opinion. In September 1990, a team of scientists from the NIH, led by French W. Anderson, performed 'gene therapy' on a three-year-old girl with a life-threatening immune disorder by inserting a normal copy of the critical gene into her white blood cells. The long and winding road towards the first gene therapy experiment on Ashanti DeSilva is colourfully reconstructed by journalist Larry Thompson, science writer for the *Washington Post* and the *San Jose Mercury News*. In his book *Correcting the Code*, Thompson lauds Anderson's perseverance and diplomatic skills, which helped him revise telephone-book-size protocols.[38] In a nutshell, Thompson's non-fiction report of the experiment shows that Anderson is as skilled in 'image doctoring' as he is in 'gene therapy' – the first being the precondition for successful implementation of the second. Image doctoring involves more than just playing to the press; Anderson continuously has to juggle to reach peer consensus, obtain permission from the Human Gene Therapy Subcommittee, and secure public opinion. Anderson's group was the first one to get the green light for gene transfer experiments on humans, where previous attempts had failed and elicited public outcry.[39]

From the very beginning, Anderson took the importance of public relations into account. As we can read in *Correcting the Code*, he is one of those scientists who 'tended to report his findings in the media before publishing them in scientific journals to assure public exposure'. During the application process, Anderson sometimes walked a tightrope between public approval and bureaucratic rejection. When he called a press conference to announce the experiment on Ashanthi DeSilva, FDA approval was not even imminent, yet he decided to keep this quiet for his colleagues and the press: 'He could use it [the press conference] to generate pressure on the FDA to approve the experiment' (19). Thompson varnishes Anderson's acts by explaining how permission for the gene transfer experiment had been repeatedly frustrated by several committees. Blaming rivalry among scientists and internal skirmishes within the FDA, he justifies Anderson's decision to withhold important information on experiments in mice from the Human Gene Therapy Subcommittee by reiterating the scientist's favourite tune: patients who are certain to die have a right to experimental treatment with gene transfer.[40] The argument that short life expectancies of patients with congenital diseases like adenosine deaminase (ADA) and severe combined immune deficiency (SCID) warrant experimental treatment with gene transfer apparently convinced both policy-makers and the public. 'The public was more willing to use the technology for human improvements than were the scientists', Thompson concludes (56).

Mainstream media seem overzealous in endorsing the promises of geneticists, as the overwhelming majority of stories uncritically herald the scientific 'breakthroughs' in genetic therapy. Gene therapy offers all the ingredients of an enthralling medical drama: heroic doctors, desperate innocent victims, a hope for cure, and last but not least, a number of hurdles that doctors have to face on their way to success. Both news magazines and non-fiction accounts like Thompson's cast Anderson as the hero, Ashanthi DeSilva as the innocent victim and gene therapy as the technological remedy against a universal enemy: congenital disease. French Anderson is portrayed not only as a hero but also as a visionary. He had started to promote gene transfer back in 1968, when other scientists considered him a loony speculator. Society simply was not ready for his ideas, Thompson asserts, yet despite colleagial contempt and against all odds he kept the dream alive. The unwarranted fear for biohazard in the 1970s clouded his dream, but he never lost his belief and trusted his imagination. Finally, in 1990, 'the time had arrived to push for people and he was excited. Anderson was no longer alone in his belief. A cadre of venture capitalists decided it was time for gene therapy to come of age' (345). Anderson's success, as we learn from Thompson, is more of a cultural breakthrough than a scientific one, because he had to pave the way in the bureaucratic jungle and remove rampant administrative scepticism. Thompson's admiration for Anderson's versatile skills in image doctoring turns this journalistic non-fiction account into a plain advertisement for genetic therapy. The excitement of this narrative leaves little room for critical assessment; in some cases of genetic euphoria, journalism has become indistinguishable from public relations.

Conversely, public relations discourse increasingly adopted the features of journalism in the 1990s. If we look at some annual reports published by biotechnology companies, their resemblance to journalistic products is eye-catching. They rival *Time* magazine not only in glossiness and layout but also in content. Genentech's 1993 annual report, for instance, titled 'The Dream Continues', contains interviews and photos of scientists and executives, profiles of the people that 'make it happen'.[41] But most striking in this report is the foregrounding of patients, who have become indispensable actors in the promotion of genetics as a drama of high hopes. Interspersed throughout the annual report, we find feature stories about patients suffering from birth defects and growth disorders, who are quoted as saying they feel much better after partaking in clinical trials, or after treatment with a Genentech product. Christopher, a ten-year-old cystic fibrosis patient, is pictured as he gayly plays around with an injection device while 'preparing his daily dose of Pulmozyme'. Another page features a child's homework writing assignment titled 'My Heroes', in which he describes how Herbert Boyer and Robert Swanson, founders of Genentech, are responsible for inventing the drug Activase, which 'saved his father's life after a heart attack'.[42] Throughout the report, Genentech's commitment to the betterment of mankind is emphasized in its columns; via the Genentech Foundation for Biomedical Science, the wonders of molecular biology are distributed through schools around the country, and Genentech's employees are said to be excited they can help 'improve the care of patients with growth disorders'. The report's features and interviews do not really differ from the *Time* magazine issue in which Collins and Anderson were profiled. Promotional discourse and journalism appear to blend in the shared enthusiasm for a once controversial scientific practice. In mass media outlets, genetics is framed less as an issue of potential ideological or political discord, and more as an unquestionable source of delight and excitement.

The images and imaginations that undergird the rationale of the Human Genome Project are simultaneously dispersed through the discourses of science, journalism, advertisement and public relations, to the extent that they become almost interchangeable. In an issue of *Nature Genetics* (1994), a computer company advertises its 'first interactive CD-Rom on human genome mapping'.[43] Reverberating Walter Gilbert's promise that the essence of a human life can be stored onto a compact disk, this ad offers the complete 'genome interactive database' at a profitable rate to laboratories and genome institutions. Images like the 'map' and the 'library' that frequently surface in journalistic accounts are recycled in advertisements. 'Quick, what is the fastest way to screen DNA libraries?' asks an ad for a screening system called 'The GeneTrapper', or, as another ad promises, 'We help put you on the map in record time'.[44] Gene sequencing machines that enable the visual display of DNA produce attractive pictures for public affairs magazines reporting on genome research, and these same image-producing qualities are abundantly advertised in specialist journals. Although the slogan 'One-Hour DNA. DNA in less than an hour. Think of the possibilities!' appears in an advertisement in *Nature Genetics*, the similarities with an ad for ordinary photo development equipment are more than coincidental.[45] Genetic screening processors are promoted in the manner of X-ray equipment or even ordinary snapshot cameras. The probability of 'seeing' one's future state of health on the basis of a 'picture' of one's current genetic predisposition, echoes the promises in magazines like *Time* of genetic 'maps' on the walls of doctors' offices. The possibility of 'taking a picture' of one's individual genetic predisposition seems no longer a mere image, as machines hit the market that actually produce such digital photographs. These advertisements, just as public affairs magazines, promise accurate assessment of one's current genetic 'condition' but also allow for a prediction of future predicaments, and implicitly urge for necessary alterations. A permanent circulation of images like 'DNA snapshots' and 'printouts' in various modes of discourse amplifies these underlying ideological messages.

The promotion of popular genome images through a variety of channels also includes cultural products. The Project's central co-ordinating office initiates and sponsors the creation of journalistic and other projects that help distribute the HGP's ideas among a general audience. *Human Genome News* regularly reports that films, documentaries and television series on the genome project are being produced with money from NCHGR.[46] Sponsoring of popular culture products is a typical phenomenon of the late 1980s and 1990s. These products invariably support the underlying ideology of the HGP, adapt its metaphors and disseminate its rhetoric. The comparison between the HGP and Columbus's journey, for instance, not only surfaces in many non-fiction accounts and documentaries, but also forms the story-line of a promotional video for the HGP.[47] The tape starts off with a comparison between the Project and the voyages of sixteenth-century European explorers: 'Imagine a map that would lead us to the richest treasure in the world. Not a treasure of jewels or gold, but a treasure far more important to humankind. This treasure is knowledge, the ability to chart our genetic blueprint'. The very same image resurfaces in an advertisement for New England Biolabs in *Nature Genetics*, in which a picture of a sixteenth-century explorer's map is superimposed on the naked body of a woman, her head facing the darkness, marking the end of charted terrain.[48] The merger of a human (female) body with an antique map, establishing the notion that the human body is the new object of a scientific exploratory mission, attaches a desirable corporate image to its producer.

A similar blending of journalism, public relations and advertisement can be noticed in the newsletter issued by the NCHGR. A newsletter is a peculiar hybrid genre, which seeks to inform and persuade. It features articles, interviews and reports, and although it does not contain any advertisements, its language is clearly promotional. Newsletters are produced by (semi-)professional organizations or by special interest lobby groups and distributed free of charge to a mixed audience of politicians, patient organizations, journalists, interested corporations and interested individuals. *Human Genome News* assumes features of journalistic discourse, but serves as a public relations instrument and a trade journal. Like a newspaper or news magazine, the newsletter updates the scientific community on important events in their circles (appointments, opportunities, a calendar of events), and reports of past events (workshops, conferences, seminars). It also informs readers of interesting publications on the HGP, yet it never lists any books critical of the Project. The newsletter does not contain any clear-cut advertisements, but names of companies and brands frequently surface.[49] Conferences are reportedly sponsored by pharmaceutical companies, and fellowships are offered through grants from biotechnology firms. *Human Genome News* increasingly resembles the glossy, sponsored magazines that have flooded the market, and which are a pivotal instrument in the process of 'image doctoring'.

The newsletter signifies the amalgamation of science, public relations, journalism and popular culture. Its peculiar format harbours and legitimizes the joint efforts to invent, produce and sell a scientific ideology, more or less pre-empting critical assessments. The way in which most images are framed and stories are presented preclude questions of a more critical nature, like how many people actually need this treatment or what the implications will be in the long run. Economic and financial considerations seem invalid arguments in stories that foreground the cure of desperately ill people, innocent victims of congenital disease. In the euphoria over human genetics, distinctions between private and public, advertisement and journalism, public relations and information, no longer seem to matter, as the discourse of *information* becomes the shared linguistic currency. Whether journalism or public relations, the popular image seems the inevitable product of an information society: a society in which images are as important as 'real' products. While public relations are still commonly labelled as activities separate from scientific practices, intent on diffusing expert knowledge to a general audience, the constant recycling of metaphors in all modes of discourse shows how fully incorporated mediating and scientific practices really are. Dissemination of genetic knowledge, in the age of genomics, derives its principal conceptual frame not from a unidirectional vector of information – the diffusion of knowledge from experts to laypersons via mediating channels – but is also more appropriately described in terms of *circulation*. Popular images of genetics, like the human genome, can be seen as equally manipulable. Stains on that polishable public face can be referred to as 'an image problem', a problem that can be countered by an appropriate public relations response. From the early stages of design, the 'doctoring' of images and information is a crucial factor to the success of gene mapping and gene therapy.

GENETHICS

The amalgamation of universities, industries and government in the 'triple helix' of genome research is paralleled by the integration of ethics in the organizational structure of the HGP. Bioethics, in the wake of emerging biotechnology industry, had carved out a firm niche for itself in the ranks of academia, and public concerns over ethics resulted in the installation of bioethics committees on various levels. Institutionally, these ethics committees and departments were located outside science proper, installed to evaluate the implications of biomedically related science and technology, and representing the interests of society *vis-à-vis* science. Until the 1990s, 'ethics' had always meant a liability in the struggle for public acceptance of genetics – a constant reminder of potential conflicts of interest between science and society. In popular representations of biotechnology stories, bioethicists often found themselves scripted in opposition to scientists. In the age of genomics, however, an avowed interest in ethics became less of a liability, and more of an asset in the image-crafting process. Ethics became a highly profiled component of genome research, and got emphatically incorporated in the HGP's organization. Yet the structural annexation of ethics into genetics also appears a valuable promotional tool.

In the age of genomics, a mounting public interest in the ethical implications of the new genetics is reflected in the large number of popular nonfiction books on ethical aspects of the HGP. Most authors agree that, while ethical questions are important, geneticists are perfectly able to tackle 'their own' responsibilities. G.J. Nossal and R.L. Coppel, in their book *Reshaping Life*, for instance, argue that molecular biologists have always displayed professional responsibility by acknowledging the ethical and moral dimensions of their work.[50] After all, did not scientists themselves call, back in 1974, for a voluntary moratorium on recombinant DNA-research? Now that genetics technology has proved entirely safe, Nossal and Coppel contend, society should leave the assessment of scientific practices to the profession itself. Genetic engineering is kept on the right moral and ethical track because it functions as 'a servant of society' and the peer group of scientists is 'thoroughly indoctrinated with locally accepted guidelines' so it will monitor any would-be dissenter (138). All types of criticism are rejected by Nossal and Coppel as Luddite emotions, serving special interests and showing irrational disillusionment with science and technology. Science itself, as these geneticists assure their readers, contains enough safety valves for it to be socially accountable. Nossal's and Coppel's understanding of ethics is that it poses endless and useless bureaucratic hurdles to scientists, which can only be overcome by fine athletes such as French Anderson. If the scientific community was left to solve ethical dilemma's by themselves, genetics would proceed smoothly towards its goal, as the authors conclude: 'All in all, we believe the world science system is in fine condition, full of all manner of self-correcting mechanisms' (154). It seems only natural that those who are able to 'correct the code' are also capable of 'self correction'.

Proponents of the ethics-genetics merger peculiarly insist on the distinction between scientists and non-scientists. Non-scientists are commonly perceived as ignorant laypersons, whose enlightenment is imminent once they have been initiated into the wonders of genetic engineering. At the same time, the populace has to be inculcated with the idea that geneticists are responsible human beings who care very much about the ethical consequences of their work. More and more scientists decry the need for government and the public sector to play an active role in the promotion of genetics. Several popular non-fiction books advocate the incorporation of ethical principles in scientific practices, and more and more scientists join the choir calling for 'genethics'.[51] A central paradox in the merger of ethics and genetics is the perceived fusion of scientific and social goals, all the while insisting on the separation between science and society. Ethics, in relation to genetics, is always interpreted as applying to the *consequences* or *implications* of science. Scientific principles, paradigms or practices themselves seem to be neutral instruments wielded by social agents but untainted by cultural or social norms. Consequently, addressing ethical or social consequences of a particular scientific project does not necessarily affect the premises or rationale that underpin that project.

The merger of ethics and genetics is also reflected in the organizational structure of the Human Genome Project. In 1990, the NCHGR established a special taskforce to examine the Ethical, Legal and Social Implications of human genome research (ELSI). In the Five-Year-Plan, the programmatic HGP document approved by Congress, ELSI is scripted as a 'safety valve', a security measure to guarantee that scientists address the consequences of their work.[52] Three per cent of the annual NCHGR's budget will be spent on research into the ethical, legal and social implications of genetics, such as potential abuse of genetic information by the state, insurance companies and employers. Many watchdogs of the HGP consider the installation of an 'ethical overhead' within the HGP a comforting thought.[53] Certainly, any attempt to address legal, social and ethical implications of genetic research should be applauded. If we look more closely at the taskforce's function, however, we cannot fail to notice a peculiar ambiguity in its aims. Though ELSI is supposed to assess the HGP's consequences critically, it is also supposed to *inform* the public about genetics research, and initiate and *promote* the development of educational materials. Implicit in this dual ambition is the idea that information is the same as critical evaluation. ELSI's professional activities comprise the strategies of public relations and journalism combined. Yet although they are responsible for a critical evaluation of the Project's consequences and promotion, ELSI studies are neither meant to assess its underlying rationale nor its scientific contentions. Conversely, HGP administrators do meddle with ELSI-related concerns.[54] The 'fusion' of genetics and ethics, apparently, is only acceptable as long as it does not touch 'science proper'.

One of the most noticeable aspects of the HGP's public relations efforts is the gendered scripting of ethics; the ELSI taskforce engenders the HGP in more than one sense of the word, providing 'genethics' with a female face. Not only does ELSI have a female acronym, it also has a female chair. Although 95 per cent of all positions within the NCHGR are filled by men, the ethics taskforce is run by a woman: Nancy Wexler, a clinical psychologist and former President of the Hereditary Disease Foundation. Her scientific pursuit, as we learn from an announcement in *Human Genome News*, is motivated by personal experience. Wexler has been involved in the quest for the Huntington's disease gene since 1968, when she learned that her mother was ill with the genetic condition, and both she and her sister were at risk of developing the disease. Nancy Wexler is eminently qualified to head the ELSI taskforce because she is both an excellent researcher and because she can relate to genetic disease from personal experience. Predictably, it is precisely her personal experience and femininity, and not her scientific expertise, that is expatiated upon in the news media. The work of ELSI, and especially the figure of Wexler, receive a disproportionate amount of media attention, capitalizing on her essential feminine qualities. *Discover*, a popular science magazine, practically turns Wexler into a saint, 'a catalyst, cheerleader, even as a mother' in the fight against Huntington's disease; *Time* and the *New Scientist* emphasize her feminine qualities, and stress how vital these qualities are to her job as

ELSI's director.[55] Heading a taskforce with an ambiguous mission, the feminine, 'motherly' figure of Wexler appears to be a great asset in the promotion of the HGP.[56]

Thrusting ethics onto the hands of women, in this case symbolized by figure head Nancy Wexler, strategically pre-empts a former detractor of genetics: feminism. Some feminists have applauded Wexler's and ELSI's contribution to the HGP as a much needed, woman-centred perspective in genetics research. They consider ethics a quintessentially feminine issue, since women are far more suited than men to imagine the social, ethical and legal consequences of science. This stance, also known as 'difference feminism', is based on the idea that women employ different forms of ethical and moral reasoning, thus counterbalancing male-centred science with an 'ethics of care'.[57] Following a line of thinking that is not at all incommensurate with the dominant HGP ideology, the script of genetics reserves specific roles for women and men in the advancement of science. As we saw previously in James Watson's gendered script in *The Double Helix*, women were assigned the distinguished task of promoters and facilitators of science. In addition, they help keep male scientists on track morally, by addressing 'genethical' implications. Women, in other words, appear the kinder, gentler antidote to male science.

Despite women's highly visible positions in the HGP organizational structure, the issue of gender is conspicuously absent on ELSI's agenda. The presence of women in the HGP may create the illusion that ethical problems generated by genomic knowledge will be properly addressed, yet this does not say anything about the way in which gender is regarded as a shaping factor in the development of that knowledge. The digitization of genetics has further disconnected body and representation, and thus effectively *un*gendered genetic corporeality.[58] In view of disembodied information, gender appears to have become an obsolete conceptual category – an uncalled for reminder of social and ideological hierarchies structuring society. Any feminism that leaves intact distinctive categories of science and society presumes that genetics can only affect women through the *implementation* of genetic practices, and fails to address the fact that the foundation of genetic thinking – the very configuration of the human body – is itself profoundly gendered. A feminism that only locates the interests of women in the *effects* of genetics research affirms the neutrality of science and technology 'proper'. It is easy to see how this shared assumption leads to the incorporation of 'feminist genethics' in the organization of the HGP.

VIRTUAL GENOMICS

Analysis of scripting and staging divulges how the preferred images, metaphors and plots stretch into the administrative-organizational structure of the Human Genome Project. The merger of various special interest and professional groups in what seems a tandem revolution in biomedicine and information technology culminates in a large network that seemingly overrides 'special interests' or specific professional responsibilities. 'Mapping' the human genome ostensibly requires 'networking': large information processing systems, but also the connection between a variety of actors and technologies involved in the Project. Networking presupposes fine-tuning of goals and methods, and most of all, (public) consent of its collective all-encompassing goal. Such ambitious goals can obviously not be attained if the execution of the Project is physically dispersed over a large number of different locales. Along with the institutional implementation of the HGP, a setting evolves that is increasingly virtual and less palpable. Due to the development of computer networks, genetic research is no longer restricted to a specific locus, but can turn into a global communal enterprise. A 'virtualization' of genomics can best be explored in the context of larger historical and cultural tendencies towards globalization.

Genomics, for one thing, provides a unique single currency: DNA-molecules, as informational units, have turned into 'data' in the merger with cyberscience. In co-ordinating the HGP, scientists now discuss 'problems of distribution, data collection, and quality control between laboratories', while talking about genetics research.[59] The joint venture between genetics and computer science, organic and digital systems, is reflected in the use of a common language and an electronically mediated discourse. Research results in genomics are increasingly published online, and time and distance are effectively abolished as significant factors in the distribution of data and ideas. Since a co-ordinated effort in genome mapping is unthinkable without a large-scale informational infrastructure, the advancement of genomics becomes contingent upon the development of the so-called information highway. In *Human Genome News* (1993), the need for a 'bioinformatics highway' to transport huge quantities of data is extensively discussed. All local operation facilities within the HGP and all potential users of bioinformation (such as clinics, laboratories, individual researchers, colleges and homes) should be linked up – a costly operation which requires 'both semantic and technical consistency among projects'.[60] A uniform infrastructure will guarantee unrestricted exchange of data among scientists and doctors, according to NCHGR officials, thus stimulating communication.

The urge for a communal transportation system, as articulated in the highway metaphor, foregrounds the necessity of universal access: everyone who can drive a car can use the bioinformatics highway – a paved route belonging in the public domain. Ironically, the need for expensive high-tech equipment substantially diminishes the number of people who can make use of these stored data. By creating a bioinformatics highway system, a completely different level of information traffic is constructed. Driving on these interstates requires a specialist licence and specialist training; consequently, it only serves and benefits those who have the skills to operate cars, understand its traffic signs and own the necessary navigation instruments to find their way around. As of 1996, all sequenced genes are made available on the World Wide Web, moderated by the NCHGR. The homepage shows a visualized body map, where one can surf from one 'chromosome' to another via simple clicks on the mouse.[61] In addition to gene sequences, one can also find current and previous issues of *Human Genome News*, and short explanations of the latest discoveries in the field. Electronic distribution of sequenced genes suggest that DNA-sequences are 'public property' which can be freely accessed and read by anyone who needs them. But just like the highway metaphor, the 'reading' metaphor fallaciously suggests that everyone who is basically literate can understand genetic data. 'Reading' the human genome always requires interpretation; the ubiquitous presence of information via the internet provides the illusion of universal access, yet, more than ever, specialized selection and interpretation skills are needed to make sense of the information overload.

A bioinformatics highway system also engenders the vision of a global community of scientists all working in 'public space' towards the ultimate humanistic goal of producing the blueprint of human life. Indeed, the control over the franchise on the representation of that space is always a powerful resource in the construction of identity and the mobilization of resources. While the global interconnected community of geneticists seems to have transgressed boundaries of space and time, in the attempt to optimize the giant mapping project, that space is increasingly less public and more commercial. While a publicly supported information system allows dispersed teams of scientists to work conjointly on the Project, effectively erasing location or time gaps, the fruits of these enterprises are harvested by multinationals who own the 'copyrights' to edited DNA-sequences. In fact, the merger of genetics and information systems renders the distinction between public and private virtually obsolete. In a culture where space – physical, geographical and virtual – is increasingly appropriated by global, commercial corporations, the boundaries between public good and private benefits seem to have vanished completely. Economic discourse has all but disappeared from the language of the Human Genome Project, as commercial stakes appear to be only 'natural'. Mergers of genome research teams with large pharmaceutical companies seem the rule rather than the exception in the 1990s, as the HGP moves closer toward its projected goal. Genomics is now at the heart of contests to reconfigure exactly what public space is and how it is occupied; digital space is post-geographical, and as yet, we lack the new concepts needed to inhabit that space.

The institutional setting of the mapping effort is frequently phrased in military and political language, which evolves along with prevailing political views. The crumbling of the Berlin Wall is presaged in the rapid expansion of genome research into former communist countries, as a result of international co-operation between research teams. In 1990, an international organization for human genome research (HUGO) was established to co-ordinate mapping activities in twenty-three countries worldwide. HUGO is nicknamed the 'United Nations of human genome mapping', emphasizing the new role of the UN as a peace-maintaining force that transcends national boundaries in its efforts to guard global peace. Scientist Stanley Cohen, commenting in *Time* magazine, squarely puts the HGP in the context of a new world order: 'We like to compete, but not on a nationalistic basis ... We intend to donate this gene map to the United Nations as a gift to the world'.[62]

HGP discourse is saturated by the rhetoric of disarmament and international co-operation; human genome research transforms from a virtual American enterprise into an international effort benefiting global peace, and becomes a trope for global political détente. This ecumenical impulse – the desire to efface boundaries and overcome difference – perfectly fits the late twentieth-century drive towards unification, whether the unification of East and West or of all European currencies. Expanding the scope of genetics research beyond national borders not only benefits its scientific goals; it conveniently meets the needs of transnational industrial corporations who have to invest (international) capital in this scientific operation. The rhetoric surrounding HUGO is not unlike a Benetton campaign; the 'global reach' of genetics research is premised on a promotional image that effectively obscures its obvious financial interests behind a screen of 'united colours'. The HGP logistics smooth over national differences for the purpose of creating a global production and consumption market. As HUGO expands the range of genome research groups worldwide, it turns global co-operation into one of its organizational pillars.

Human genome euphoria, as we have seen in this chapter, thrives on the merger of a number of dichotomies – reality versus representation, organic versus digital, genetics versus ethics, public versus private – whose hierarchies remain paradoxically untouched. The book and the map, in their digital manifestations, no longer serve as metaphors, yet we hang on to their representational status and conventional interpretations. Images and plots of great discoveries and map-making expeditions by far outnumber more appropriate images of circulatory networks. Information and genomic technologies permit visual access to inner parts of the body, while the technologies themselves remain virtually 'unseen'. Ethics, in the HGP, is incorporated in genetics, inferring the notion that science is fully immersed in society; although 'genethicists' are supposed to question the implications of genetics, they leave the assumptions of 'science proper' untouched. And the emergence of a global, virtual space seemingly overrides private and commercial interest in the communal striving towards a unifying goal. While the implosion of categories is complete, genetic bliss cannot succeed without an insistence on their distinctiveness; to remain effective, the fusion of empirical and ontological zones necessarily adheres to the very hierarchy it seemingly undercuts.

The critical power of digigenetic euphoria lies precisely in this double language; enthusiasts mobilize science at the heart of social relationships by invoking 'natural' dimorphisms and commonsense images. In staging and scripting the public meaning of genomics, distinctions between journalism and public relations, advertisement and popular culture, also notably fade. Dissemination of knowledge is increasingly attained through a careful circulation of images and ideas among a number of dispersed discourses. Responsibility over the creation and distribution of images, previously claimed by particular professional groups, has definitely been reallocated to all participants in the staged performance. 'Image doctoring' does not remain on the periphery, but moves to the core of what constitutes the Human Genome Project. The material, scientific object of genetics research as well as its popular images now seem to share a common origin in the language of information.

The border crossings apparent in the reconfiguration of genomics interrogate concepts of nature along with concepts of science, and questions epistemology along with representation. As we shall see in the next chapter, critical assessments have not been locked out of public discourse on genetics. Throughout the period that was dominated by genetic euphoria, there is a persistent strain of 'biocriticism' that probes the scientific, cultural and ideological assumptions underlying the 'big genetics' typified by the Human Genome Project. This 'biocriticism' is actually characterized by its very insistence on the *inseparability* of science from culture, science from non-science, and science from the representation thereof.

6 Biocriticism and Beyond

Images of biophoria are dispersed over a variety of discourses, recycled through science, journalism, public relations, advertisement and popular culture. Given the great proliferation of favourable genome stories, one could easily conclude that genomics' advocates have no opponents, or that all previous antagonists have joined in a gleeful celebration of a communal scientific project. Yet in the age of genomics, biocriticism is poignantly present, although it emerges neither as a unified counterdiscourse, nor as a 'special interest' voice that defines itself squarely in opposition to genetics. Instead, critical assessments of genome research are equally dispersed through a variety of discursive forms, from academic essays to science fiction. A number of scientists, journalists, artists and writers question the hegemony of prevailing gene metaphors, subvert genetics' preferred plots and suggest alternative images and imaginations. They carefully deconstruct popular images of the human genome, yet re-imagination requires more than the deconstruction of gene images; it also involves a reconceptualization of nature, culture and technology.

The accuracy and validity of popular genome metaphors have been scrutinized by several biologists and specialists in related fields, who refuse to take the vernacular on the human genome for granted as a mere tool in the distribution of genetic knowledge. Popular images and metaphors of the genome are regarded as efficacious resources in the creation of public knowledge. Biologist Thomas Fogle, in his elucidating article on the use of information metaphors in the Human Genome Project, states that geneticists have a 'special responsibility to communicate clearly with a wider audience about the nature and significance of genetic information' and sets out to explain why metaphors about DNA programs and blueprints are 'neither biologically realistic nor socially acceptable'.[1] The motivation for crafting powerful symbolic language is apparently to facilitate public understanding and acceptance of a principally abstract technical enterprise, but Fogle lists several reasons why these images are dangerously misleading.

First, the comparison of DNA with blueprints and programs engenders the interpretation of genetics as a matching process between a single gene and a trait. This imposes a sense of biological determinism onto what is basically a contingent relationship between biochemical pathways, cellular structures and physiological processes. Fogle claims that the prevalent metaphors do not coincide with what is discovered at the molecular level. In the public mind, DNA has come to connote a beanbag collection of genes whose expression determines traits. Most images impart a static relationship between phenotype and genotype. In the dominant computer program analogy, all information in the phenotype is already encoded in the genotype, yet cellular machinery itself does not contribute program information. Fogle objects to the inflexibility of this model: it misrepresents development's role in the expression of traits. DNA, he argues, does not just mechanically yield products in a passive cellular environment, it also changes its actions depending upon that environment. DNA is not a set of absolute, all-containing orders; cell and DNA are constantly interacting, mutually affecting each other's change. Biological 'programs' only make sense in the context of the system of which they are part:

> DNA acts in the dual capacity of program and data, and the cellular machinery likewise acts as both passive interpreter and program. The genotype and phenotype are intertwined, both acting responsively to the other, both contributing to the process and the result (540).

Fogle not only points at the inaccuracy of prevailing metaphors, but also finds them delusive. They remain popular with the public in part because they fit a research model that separates genetic from environmental contribution or nature from nurture. The heavily publicized 'search for genes' provides a major selling-point of the HGP, but it narrows the public's vision to a highly limited aspect of the relationship between genotypes and phenotypes. Coding portions of the genome is often most valuable for medical applications; it also offers the opportunity to researchers and companies to tie 'function' to 'sequence' – clinical correlations between specific genes and specific traits which can be highly profitable because they are uniquely patentable and 'ownable'. But this narrow focus does nothing to further the understanding of the complex interrelations between DNA information and cellular information. A final objection to the popular images is that they attach a strong symbolic and metaphysical meaning to the genome, which renders the grander claims of the HGP – a blueprint to the understanding of human life – highly inflated and conspicuous. Geneticists only study a few relationships within a vast network of variables, but the HGP is not 'the answer to what makes us human, in part because the biological network is in continuous flux' (545). Fogle concludes that the biological and medical community needs to put the HGP in a perspective broader than the narrow search for the 'map'; they should be more sensitive to the metaphors they choose to inform a general audience.

Scientists and academics critical of the HGP seem particularly concerned about the effect of popular images and rhetoric on the conceptualization of scientific knowledge. Historically, the choice of metaphors in molecular biology reveals a mutual shaping of scientific concepts and their popular semantic coatings. In his analysis of the work of early geneticists, Richard Doyle shows that the crux of the 'new' genetics lays in a shift from metaphors of the phenotype to metaphors of the genotype.[2] Erwin Schrödinger's coining of the code metaphor, emphasizing the notion of 'pattern' or an underlying set of rules, shifted biologists' attention from the individual or the phenotype to the genotype – an epistemological shift which must be seen as a 'fundamental reprogramming of the rhetorical "software" of genetics, and, by extension, molecular biology' (55). The rhetoric of the genotype places all power within the code, and none within the development of the organism. Later theories, whether Francis Crick's Central Dogma or Richard Dawkins's views of bodies as 'survival environments' for selfish genes, can be traced back to Schrödinger's substitution of phenotype metaphors with the language of the genotypical 'code-scripts'. Although Doyle essentially shares Fogle's conclusion about the mismatch between scientific reality and its popular metaphors, he also argues that 'rhetorical software' and 'genomic hardware' shape each other. Scientists are constantly engaging in acts of *translation* rather than explanation.

It is precisely the intricate connection between scientific concepts and popular representations that accounts for the wide public support of genomics' grander claims. The life sciences, according to Doyle, are made up of scientific techniques as well as rhetorical ones. Schrödinger's metaphorical substitution enabled Watson, Crick and others to equate 'life' with the structure of DNA. Doyle dissects how Schrödinger's rhetorical move made it possible for George Gamow in 1954 – and later for Francis Crick – to suggest a conceptual model for the DNA-protein relation that helps translate DNA to 'life'. Through diagrams and texts, Gamow argues that amino-acids 'get caught into the "holes" of deoxyribonucleic acid molecules' (63). Using a lock-and-key metaphor to explain the relation between DNA and protein, he provides a conceptual bridge to view genes as the 'essence' of life. Gamow embroiders on the image of DNA as a decodable text, the secret of life that is decipherable and knowable. The implosion of life and molecules, engendered by the images and metaphors launched in 1950s genetic theories, strongly reverberates in the inflated grand claims of human genome researchers. The metaphor of DNA as language, of the genome as a Book of Life, promises the possibility of a final solution to the uncertainties of living. Not surprisingly, it is precisely this grander claim of the scientific narrative that appeals to a general audience, and accounts for the growing popularity of genetic theory. Metaphors and images, as Doyle concludes, are 'vital' to the genetic enterprise because they interlock scientific, cultural and theological narratives.

Another profound critique of the content and impact of popular representations stems from feminist biologists. The dominant configuration of the gene and the genome, as Bonnie Spanier contends, imposes a paradigmatic hierarchy in molecular biology.[3] The centrality of the gene, and its importance as the 'controlling molecule of life' places genes at the top of the micro-cellular hierarchy, in control of all cells. In this model, the nucleus is often figured as male, controlling female cytoplasm. Classical genetics is thus based on the primacy of nuclear (male) heredity and a subordination of cytoplasmic (female) heredity. The new biology has elevated what is associated with males – nucleus, plasmid, DNA – to a superior status, while diminishing extra and intercellular influences. Common descriptions of the gene and the genome, Spanier argues, are not so much inaccurate as they are partial, and hence misleading:

> In a nonhierarchical reconstruction, genes are only one part of a multi-faceted, multidirectional hologram of "life." The term "gene product," as currently used, narrows the concepts to focus on the importance of DNA above the other components and processes of the cell and organism. The result of this narrow scope of "life" is a misrepresentation, a one-dimensional cause-and-effect process that leaves out biological history and the context in which DNA exists and operates (93).

In line with Fogle's and Keller's criticism, Spanier argues that prevalent scientific images of the gene enforce a conceptual separation of genetic from non-genetic interactions, parallelling the nature-nurture dichotomy.

Cultural beliefs embedded in molecular genetics often remain implicit in scientific discourse, but appear explicitly in their popular reincarnations. Much of feminist criticism pertains to popular gene images that appeal to basic masculine fantasies of penetration, appropriation and conquest. The 'book' and the 'map', according to Mary Rosner and T.R. Johnson, are variations on ancient parables of the power of science and the mastery of nature.[4] They argue that the concept of 'reading', in the sense of unilateral deciphering, is a 'brutally coercive device', because it conflicts with the values of 'multiplicity, fluidity, and indeterminacy that many associate with feminism' (109). The digital genome elicits concepts of disembodied body-parts which can be probed and reassembled without recognizing the genderedness of the human body. Images of maps impart assumptions of 'virgin' land to be discovered and conquered by male hunters and explorers. To counteract images based on binary frameworks, the authors suggest introducing new images and metaphors that blur the line between 'the figurative and the literal or between science fiction and social reality' (125).[5]

Human genome sceptics direct their criticism at the images that encode and sustain research practices, and at the way geneticists disseminate knowledge to a general audience. Neither of these contestants of genomic claims position themselves 'outside' or 'in opposition to' genomics, nor do they assume that science is mere rhetoric, or that rhetoric is the main culprit of the determinist fallacy. Instead, they point at the inadequacy and inaccuracy of dominant images, and recognize popular sources – next to scientific sources – as significant loci of 'geneticization'. Although neither of these critics provide viable alternatives to the pervasive images of 'life', their insights are critical to the understanding of possible re-imaginations of conventional plots.

REDESIGNING 'LIFE'

In addition to revealing the fallacies hidden in prevailing metaphors, some biologists and artists make an attempt to subvert worn-out or delusive genome plots and creatively redress ingrained meanings of these fixed images. Molecular biologist Robert Pollack and novelist Richard Powers both take on the 'language' and 'code' metaphors to question the monolithic contentions of the Human Genome Project by stretching the attenuated image of language into the expansive realm of literature. They also rewrite traditional plots used in popular genome narratives: plots of libraries, treasure hunts, famous literary tales and stories of double helixes. Both authors try to abridge the assumed gap between science and culture, as they equally draw from literary and scientific sources to express their discontent with prevalent images of genome euphoria.

In *Signs of Life*, Robert Pollack severely criticizes the common configuration of the genetic structure as a language – an image that defines language as a static set of coding rules.[6] Instead of taking linguistic concepts as a model, he proposes transferring the analytical apparatus of literary criticism onto genetics. To view genetics as literature rather than language allows recognition of the complexity of DNA. Taking advantage of the elaborate set of meanings this expansion brings along, Pollack tempers unmitigated biophoria and takes on the Human Genome Project's underlying determinist assumptions. In line with prevailing metaphorical preferences, he describes DNA as a 'long, skinny assembly of atoms, similar in function, if not form, to the letters of a book' (5). His notion of reading, however, involves more than deciphering: 'The cells of our bodies do extract a multiplicity of meanings from the DNA text inside them, and we have begun to read a cell's DNA in ways even more subtle than a cell can do' (5). The inherent polysemy of DNA texture enables multiple interpretations of the same letters; this infinite potential of meanings generated by a DNA text should be acknowledged by geneticists as a great asset rather than a disadvantage. Reading, according to Pollack, is an activity not directed at intellectual closure, but geared towards intellectual exposure. Therefore, the scientist suggests that geneticists borrow the analytical apparatus from literary theorists to raise a different set of questions: Are cells reading or are they merely decoding? Will there be a 'canon' of DNA texts? What can we learn about the language these DNAs share, the dialects they produce, and the arguments they articulate? The discourse of literary criticism, in other words, propels questions of polysemy and multi-interpretability.

Pollack pursues his interdisciplinary reconnaissance by disassembling some of the most exploited images in genome sequencing, such as the 'library', 'translation' and the 'Book of Life', to put a twist to their dominant meanings. Explaining the structure and function of the twenty-three pairs of chromosomes, he likens them to the two replicating towers of the World Trade Center ('an odd sort of library') where each stack holds a duplicate set of books. Library staff take down one volume, photocopy an article and put it back on the shelves. Human mistakes in the process of copying and reshelving are only as common as genetic mistakes, which we have come to know as mutations. The author of *Signs of Life* proposes reading human DNA as 'a work of literature, a great historical text', but he immediately points out that such text can never have a single meaning; the human genome is not a 'sacred text' with only one valid version. The value of literature lies in its potential to generate endless meanings – meanings that are contingent on the specific (historical) context of its readers. Comparing three translations of a paragraph from the New Testament, taken from different historical periods, Pollack shows that each version yields a profoundly different interpretation of the same words. Transposed onto genetics, this implies that meanings will multiply with time, and no single allele is ever going to be the sole 'correct' version. The idea that there is a single, canonical interpretation of the human genome, whose precise alleles we might hold up as a perfect mirror, is not an innocent misconception but a dangerous tenet shared only by fundamentalists. The purpose of expanding the language metaphor into the realm of literature is to show that DNA opens up whole new worlds of interpretation; the leap from DNA to protein is as arbitrary as the relation between signifier and signified.

Pollack further deconstructs the concept of 'translation' to unnerve popular comparisons of the human genome to historical events. Geneticists often liken their goal to 'crack the code' to deciphering the Rosetta Stone, the code of which was 'broken' by Champollion in 1821. But the word 'translation', in common usage, is as much a misnomer in the context of the historical discovery as it is in relation to genetics. The Rosetta Stone was a rebus, a hieroglyphic sculpture that carried different language inscriptions of the same text. When geneticists talk about 'translating' the human genome they talk about transposing a three-dimensional language (DNA) onto a linear representation of alphabetic units. Translation thus intrinsically entails a change of meanings; just as in ordinary translation, a word never means the same in another linguistic or representational context. Deciphering the 'meaning of life', as Pollack argues, is the production of a 'consensus trans*literation*' but never a translation (90). A transliteration of a string of four letters is no more likely to reveal multiplicity in a gene than 'the transliteration of a poem by Pushkin from Cyrillic into English would enable an English speaking person to see layers of meaning in a poem' (12).

Technological upgrading of the language metaphor into digital information offers a new and exciting view of the possibilities of genetics. Once we have ordered and stored library fragments, we can start to edit existing genes. Pollack's introduction of the 'molecular word processor' signals the potential of gene therapy as the 'revision' or 're-authoring' of life:

> To use the molecular word processor is to write in the language of DNA freed from the constraints that natural selection has placed on the possible meanings a cell can put into or draw from its genomic texts. With it we can insert or read new meanings into existing genes, and construct new contexts in which gene sentences can take on new meanings ... As our tiny vocabulary of protein meanings grows, we will inevitably be drawn to the idea of a molecular literature entirely of our own creation (96).

Gene manipulation ('the recycling of microbal swords into molecular plowshares') puts scientists in the position of authors. Pollack evaluates the prospect that humankind can edify its own book of life with mixed feelings. Transgenic mice already populate the laboratories, and experiments with transgenic people may not be far away. 'Transgenic', however, does not necessarily imply the creation of monstrous hybrids; a child born without a congenital disease passed on by its parents is also a transgenic product. Pollack's point is that mutations are never predictable, so the consequences of 'improving' the genetic texture are impossible to calculate: 'Single genes may one day be totally understood, but the overall meaning of the genome will not be a predictable sum or product of these separate meanings' (152).

The very indeterminacy of the human genome should seduce scientists to see DNA as literature, rather than as language, and to treat its texts accordingly; they should abandon the hope for intellectual closure and rejoice in the pleasure of endless, multiple interpretations. By transplanting the apparatus of literary theory onto the life sciences, Pollack assumes that the narrow view of scientists will be expanded by the humanities' perspectives. Molecular biology should be appreciated as a creative art as well as a science. Through their ability to edit and create the 'texts of life', authors and readers may give whole new meanings to the human body. A precondition for the edification of the 'book of life' is that its authors start to re-imagine the 'fixed' plots that are passed down to us. Literary history has given us commonplaces of imagination like *Faust*, *Frankenstein* and *Brave New World,* but literary texts are much richer than their conventional interpretations. Literature always provides endless resources for re-imagination (180–81). If scientists adopt the multifarious concept of literature, they may begin to appreciate the variations rather than the fixed code.

Adopting the popular metaphor of genetics as language, but expanding its figurative range to include literature, Pollack forges an ideological and conceptual shift in dominant genetic thinking. He reclaims analytical frameworks common in the humanities and reappropriates them for a double purpose: to change the conceptual model prevalent among geneticists, and to explain a highly abstract scientific theory to non-scientists. His tactic seems aimed at bridging the gap between the two cultures and at criticizing the essentialist rationale underlying the HGP. Pollack deconstructs for us, by means of analogy, genetics *in terms of* literature, and hence serves as an 'interface' between the seemingly incommensurable worlds of science and culture. Although he proves himself a staunch critic of the reductionist tendency in genomics research, he also presumes a particular view of culture – a collection of precious books laid down in a canon of texts – and projects that onto science: 'Once we finally see that all genomes are a form of literature, we will be able to approach them properly, as a library of the most ancient, precious and deeply important books, only then can the new biology be born' (177). Besides viewing culture as a canon of texts, he also assumes a self-evident split between representation and reality. The idea that metaphors are representational models that do not really affect scientific reality, for instance, is evident from his argument that the advance of computer science has not really changed our concept of how DNA works. The word processor and notion of information processing serve to illustrate the intricacy of genomics. Yet he never addresses the fact that the digitization of genome research is actually inscribed in its practices, and that information processing is no longer a mere figure of speech, but crucial materiality in genome research. The power of polysemy – the indeterminacy of the human genome – is at the heart of Pollack's critical message, but the actual form of *Signs of Life* seems incongruent with its basic message. Science can only be explained in terms of literature, and while the gap between these two cultures can and should be bridged, there seems a vector of meaning transfer but no sign of reciprocity. Pollack unidirectionally projects his interpretation of culture onto science; we can read genetics *as* language or literature, but what counts as metaphorical knowledge in one field can never transcend its ornamental function and become a shaping factor of scientific materiality.

Whereas Pollack borrows the conceptual framework of literary criticism to explain genetics, Richard Powers carries the conceptual frameworks of genetics over into literature. In *The Gold Bug Variations*, the inherent polysemy of language and the human genome is reflected in its intricate narrative plotting.[7] Simultaneously, the novel subverts some of the fixed plots usually associated with genetics. In its title, the novel reflects the intertwining of three different disciplinary fields: music, science and literature. *The Gold Bug Variations* is modelled after Bach's *Goldberg Variations*. The thirty chapters of the book, framed by an opening and closing aria, exactly correspond to the format of Bach's famous composition.[8] All of Bach's variations derive from the same basic notes, as they are all variations on a theme. Interwoven in this design, Powers's narrative plotting borrows its structure from a scientific model: the double helix.

Two love stories, one set in 1957, the other in 1983, converge to constitute a double stereo effect. Stuart Ressler, a geneticist who, in 1957, is 'one of a new breed who will help uncover the formula of life', falls in love with his colleague Jeanette Koss. Their team of molecular biologists sets out to crack the genetic code, a few years after Watson and Crick have discovered the helical structure. Gradually, Ressler comes to understand that the 'secret of life' will never be cracked unless scientists choose a different way of looking at the coding problem. Listening to Bach's *Goldberg Variations*, the young geneticist realizes that the core of the problem is not finding the universal genetic code, but acknowledging its endless potential variations. Twenty-five years later, Stuart Ressler, having quit a promising scientific career, works at Manhattan Online (MOL), a giant data-processing company where he shares night shifts with Todd Franklin, a graduate student in art history. Intrigued by the geneticist's peculiar career change, Franklin tries to reconstruct the facts of Ressler's life. While searching for information in the library, Todd gets help from librarian Jan O'Deigh. In their joint effort to collect and put together the details of the molecular biologist's life, the two become romantically involved. The two love stories, separated twenty-five years in time, spiral into a double helix – the two strands braided together by Jan O'Deigh, who attempts to reassemble Ressler's *vita*. O'Deigh's reconstructive journey forces her to look simultaneously for biographical facts and scientific theories, as the key to the secret of Ressler's life is ultimately locked in the genetic 'secret of life'.

The Gold Bug Variations is thematically composed around the three metaphors prevalent in genetics: language, the library and computer processing. The genetic code, in Powers's novel, is not *like* language: it *is* language, but language in all its facets, including polysemy – its intrinsic potential to generate infinite meanings in different compositions. The mistake most scientists make is that they are looking for the underlying rules of language formation, whereas the essence of language lies in its various enunciations. To elucidate this point, Powers reverts to Edgar Allan Poe's story *The Gold Bug. The Gold Bug Variations* subverts the traditional treasure-hunt plot, which is cogently summarized by Jeanette Koss: 'simple letter frequency and word pattern trick leads scholar to pirate's treasure' (76). Most scientists take a conventional approach to the coding problem: if they only decipher the riddle, the decoded message will lead them to the treasure. Yet Ressler realizes the genetic game is immensely bigger and the stakes are higher:

> The treasure in Poe's tale is not the buried gold but the cryptographer's flicker of insight, the trick, the linguistic key to unlocking not just the map at hand but any secret writing. ... Not the limited game of translation but the game rules themselves (77).

Ressler provides this key when he finds a way of reconceptualizing the coding problem as a problem of musical theory. The essence of music is not the standardized sign system, inscribed in notes and staffs, but the infinite variations generated by these notes; subsequently, he exchanges the biological paradigm for the theoretical apparatus of music. As Ressler listens to Bach's *Goldberg Variations*, the music triggers a philosophical discernment that changes Ressler's career: 'No wonder this Bach fella is so great a composer; he anticipates Watson and Crick by two hundred years' (191). For the outside world, the geneticist quits a promising career in science, though Ressler himself is convinced he has merely changed his direction of thinking about molecular biology.

The parallels between different coding systems induce a meta-reflection on the notion and function of metaphors. The genetic code, as Ressler becomes aware, is itself a figure, a metaphor, 'the code exists only as a coded organism' (271). Empirical research and representational theory, in this way of reasoning, have collapsed beyond distinction. When Jan O'Deigh, decades later, tries to reconstruct Ressler's scientific argumentation by studying the basics of both genetics and musical theory, she realizes that almost her entire understanding of science depends on the use of figurative language:

> It hurts to discover how much my understanding relies on analogy, pale figurative speech. ... Scientific method itself – from diagrams to symbolic formulae to phenomenal descriptions – relies on seeing things in reflected terms. Will I ever get it? Code is itself a metaphor (166).

The 'code of codes', the human genome, fails in its representation of reality, yet we lack another instrument to communicate its meaning or call it into existence. By the same token, it is impossible to transmit one system of knowledge into another without losing some of its meaning in the process – translation can never be the same as transliteration. Translation of the genetic code is unattainable because the human genome defies univocal interpretation: 'The closest he would ever get is simile, literature in translation, the thing by another name, and never what the tag stood for' (516). Representation, in other words, is always distortion, yet at the same time it constitutes reality: the 'code' – musical or genetic – is not merely a representation of the human genome, but its very condition.

The library plays a crucial role in *The Gold Bug Variations*, both as an image and as a setting. Jan O'Deigh, the female protagonist, is a librarian – an archivist searching for an algorithm in the labyrinth of information. As a professional navigator in the ominous flow of facts and numbers, O'Deigh observes that information is never quite the same as knowledge. Librarians are also interpreters: they have the power to steer information, to retrieve and reassemble facts of life from obscurity. O'Deigh conscientiously manages the reference desk and the 'Question and Answer board', where she pins up a daily notecard highlighting a long forgotten fact in history. The library, as a symbolic space, is a mental labyrinth that contains an endless number of possible routes to enter and exit. A map of such a place, even though it would help you find your way through the labyrinth, will never adequately represent the multiple potential routes: 'The map is never quite the place, nor the place as navigable as its image' (368). Along the same lines, a map of the human genome, despite its projected potential to mirror the sequences of DNA-strings exactly, will never reflect the varieties of 'life' that may come out of this basic structure. Geneticist Ressler is predictably sceptical about the value of the Human Genome Project: 'Even the complete library, unattainable, will never begin to hint at the books, the stories the string might have produced' (606).

In her attempt to reconstruct Ressler's scientific catharsis, his epistemic shift in genetic thinking, Jan O'Deigh ties in his new conceptual frame with her own experientially based frame of reference, when she compares the genome to a huge library: 'If this business is a business at all, it must be a lending library – huge, conglomerate, multinational, underfunded, overinvested. ... The word I'm looking for, the language of life, is circulation' (326). Circulation is the metaphor that ties in a conceptual model for the genome with a model for the structure of knowledge: the contingency and interaction between producers, receivers and resources of knowledge mirrors the structure of molecular information. It is indeed the circulation of ideas from a variety of representational systems – music, literature, science – that insures the generation of new knowledge, yet this flow is never predictable. *The Gold Bug Variations* reflects the contagiousness and creative potential of the mutual infection of disciplinary ideas.

Besides deconstructing the meaning of 'codes' – musical, genetic, linguistic – the novel also examines the implication of an electronic update of these metaphors. The translation of 'life' into digits has not only transformed genetics, but digits have become the universal currency in a postmodern world which capitalizes on processable information. Powers's novel shows how people's physical and social conditions and desires are digitally coded in order to be categorized and quantified. One of Ressler's friends, for instance, works for the Social Service, where she processes job applications by attaching a numerical code to applicants with risky health conditions. She knows by heart the codes for all types of illness – the ultimate digital reduction of human beings. The giant knot of digital information that encodes all of people's financial, physical and personal data, is symbolically represented by Manhattan Online (MOL), the data-processing firm where Stuart Ressler and Todd Franklin work night shifts to keep an eye on the flow of billions of bits and bytes.

The fact that voluminous streams of data do not just represent, but actually define people's lives in the 'real' world, is clarified by two telling events. Ressler and Franklin witness that one temporary hitch in the computer system one day causes financial markets to fluctuate the next day, as digital subsystems are all interconnected. And when the two colleagues tinker with the processing system in order to procure their co-worker Jimmy a financial bonus, he ends up losing his health insurance, due to an unforeseen mutation in the digital order. Two days later, Jimmy suffers a stroke and the two perpetrators of this computer crime desperately try to locate the processing mistake to get him back on coverage. Illness and health are regulated by the computer, not only through genetic inscription but also through their interconnectedness with other digital systems, such as insurance and banking. Digital systems show the same 'bugs', or unpredictable mutations, as organic systems. Life, in all its manifestations, is regulated through 'the absurdity of a language that made oncology and ontology differ by a single mutation' (636). The reciprocity between technologies of representation and archaeologies of information is ubiquitous; genetic algorithms are no longer mere registrations of the organic, but the representational system itself has substituted the biological system it once signified. Genetics and digital information systems have imploded, and terms like 'virus', 'bugs' and 'evolution' now equally apply to both. Form, structure and images of *The Gold Bug Variations* upset the modernist hierarchy between empirical grounding and representation; the digital code is not a representation of the genome but has come to function as an autonomous signifier in another context.

The story defies strict genre boundaries just as the genome does not lead to a definition of 'life'. In the end, the entire novel turns out to be a vain attempt by Jan O'Deigh and Todd Franklin to write Ressler's biography. While O'Deigh focuses on his scientific and intellectual development, Franklin concentrates on his personal life. Both 'editors' of life realize that a biography is a heterogeneous knot of ideas, theories, facts, images and experience, not only Ressler's but also their own and others'. Beyond the biographical level, the fallacy of 'writing life' is reflected in the novel's juxtaposition of discourses, resulting in a Bakhtinian dialogic imagination: scientific theories – Jacob's and Monod's messenger theory for instance – are paired off with literary theories and metaphors to the extent that makes it hard to distinguish science from fiction. Textual features that signal distinctive discourses disappear beyond recognition; allusions to scientific publications – Watson's and Crick's famous *Nature* article – poems or theories acquire meaning only if one is alerted to the possibility of endless connections via extratextual and intertextual references. Readers of *The Gold Bug Variations* are forced to act as librarians, always deriving their own knowledge from basic information, producing their own compositions from the same basic notes.

Whereas Robert Pollack's *Signs of Life* urges a unilateral transposition of literary theory onto genetics, Richard Powers's *The Gold Bug*

Variations invites border crossings in both directions. The novel creates a discursive universe in which science, fiction, biography and real life reverberate in the productive circulation of ideas. Empirical problems become problems of representation and vice versa. The two cultures, which Pollack tries to interface in the humanistic tradition, have simply ceased to exist in Powers's novel. Knowledge, as the structure and content of *The Gold Bug Variations* imply, is derived from the infinite concatenation of signs, ideas, methods, models and theories from all disciplines. Circulation and variation seem the proper metaphors for both genomics and the dissemination of genetic knowledge into culture – a technoculture in which image and referent, genes and bytes, organic and digital systems, compete within the same domain, and where both body and image are subject to the vagaries of narrative manipulation.

UNRAVELLING GENE MYTHS

A common denominator in many criticisms of human genome euphoria is that the 'gene' and the 'genome' become crystallized into things, rather than being fleshed out as complex processes. Reified as entities, they start to function in society as signs with fixed meanings, only to become signifiers in other contexts. The gene, the genome, the helical model and the map are increasingly detached from the biological discourse in which they arose as metaphors; they are recycled in a variety of social and cultural contexts to the extent that they become accepted as models for both cultural and social practices. At the heart of a critical assessment of genomics is a decontextualization of scientific images. Several authors of non-fiction books have scrutinized specifically the popularization of genome images in relation to contemporary social and cultural practices. Dorothy Nelkin and Susan Lindee in *The DNA Mystique* and Ruth Hubbard and Elijah Wald in *Exploding the Gene Myth* depart from the assumption that popular culture is a crucial medium for the dissemination of scientific knowledge, as genetic principles rapidly find their way into other domains of public life. Scientific images – maps, models, metaphors – provide scripts for social practices such as politics, law or religion. But the process of dissemination is as much the object of critical scrutiny as the outcome; Hubbard particularly pays attention to the role of journalists, who are more than 'mediators' in the image-making process.

Nelkin and Lindee present their work as an 'analysis of folklore', assuming that popular images and stories of genetics are not isolated (arte)facts, but that they reflect and convey the message of genetic essentialism.[9] Their purpose in exposing the gene as a cultural icon and genetics as a mythical narrative is not so much to identify distortions of science, or debunk scientific theories, but to show how scientific and popular presentations intersect in shaping the cultural meaning of the gene: 'The precise scientific legitimacy of any image is less important than the cultural use that is made of it' (4). In their study of the changing constellations of meaning, they first and foremost try to understand how these images spread, and how they solidify prevailing assumptions about health, disease, normalcy and deviance. The venue these authors take differs in its basic approach from studies that prioritize the study of ethical, legal or social implications of genome research. Discriminatory policies or judicial decisions are not so much a *consequence* of scientific theories and practices; the potential for these consequences is already prefigured in the very formulation of genetic theories.

One area in which the idea of genetic predisposition has clearly proliferated as an uncontested truth is family law. Inscribed DNA-sequences are looked upon as evidence in legal disputes over parenthood. Biological parents enjoy increasing legal preference over social bonds established by parents who raised a child without delivering the genetic material. Precedent-setting court cases have changed the definition of 'mother' from the woman who delivers the baby to the woman who delivers the genetic material.[10] The technical possibility of identifying one's DNA may be regarded as a scientific practice, the implementation of which is fully regulated by social and legal institutions; yet there is no denial that court officials unconditionally accept genetic evidence when they define family bonds in terms of molecular inheritance. A similar example of unconditional acceptance of genetic thinking can be found in the insurance industry, where genetic forecasting tends to be elevated to the level of future fact. Nelkin and Lindee extensively illustrate how actuarial thinking is not so much a specific *implication* of genetic theory, but that the widespread acceptability of this concept is ingrained in our cultural beliefs about the predictability of future health on the basis of genetic predisposition.

The DNA-print, in other words, has become the ultimate identifier, establishing the essence and identity of a human body. In the previous chapter, we have seen how the 'identifier' materializes in advertisements for DNA-sequencing machines. DNA imprints start to function as documentary proof of health or identity, like a biological social security card. Nelkin and Lindee show how the gene is figured not just as a biological structure, but as a cultural icon, a symbol and even a magical force. The gene, in quotidian conversation, is often used as an explanandum for disease, a scapegoat for deviance or a legitimation for inequality. Genes, according to the authors, have become reified 'products' that shape our narratives of social, sexual or racial difference. As a deterministic force, the gene has transformed public discourse about the source of social problems and the exoneration of guilt. 'Bad genes' can both account for, and absolve one from, criminal behaviour. Yet genetic narratives, Nelkin and Lindee observe, do not inherently oppress or liberate; these narratives are only problematic in a 'society that tends to overstate the powers of the gene' (126).

Over the years, the gene and the double helix have mutated into cultural icons with petrified meanings. A frequently heard statement like 'it's my DNA' or 'it's in the genes' reveals how a scientific principle informs a platitude. The double helix – once a model for the structure of DNA – surfaces in mass media as a self-explanatory icon of genetic determinism. Covers of magazines feature scientists wrapped in double helical strings as politicians wrap themselves in the flag. The two spiralling base-chains have come to be commonly associated with medical progress, signalling universal hope for curing congenital and other disease. Biotechnology companies feature the double helix in their logo, just as doctors and pharmacists use the staff of Aesculapius as a symbol of their commitment to humanity's health. On a connotative level, the double helix appears to convey the idea that, although we may be flawed by our genes, the medical-biotechnology conglomerate will go to the roots of disease. Like the flag or Aesculapius's staff, the double helix has come to signify a professional, ideological and political conviction. Arcane symbol of a once revolutionary theory, it metamorphosed from an analog model into a cultural icon of a hegemonic belief.

Genetics has indeed proliferated, on a cultural level, as a 'mystique', as Nelkin and Lindee contend: 'Instead of a string of purines and pyramidines, it has become the essence of identity and the source of social difference' (198). Metaphors, models and images that were once introduced to exemplify the working of DNA now form the root metaphors and conceptual archetypes that sustain the myth of genetic determinism. Nelkin's and Lindee's goal to untangle the mythology of the gene is akin to Ruth Hubbard's attempt to 'explode the gene myth'.[11] Hubbard's book is the critical counterpart of popular non-fiction accounts written by scientists who herald the Genome Project. She sets out to bare the roots of popular myths by analysing a number of newspaper stories on genetic discoveries. Journalists and scientists, she contends, are united in their embrace of genetics and biotechnology as the wave of the future, and it is important to examine the framing of this unarticulated consensus – the loci from and ways in which these beliefs are distributed.

Hubbard draws particular attention to the many 'gene-for-everyhing' stories expatiated upon in the press, as news stories uncritically adopt scientists' claims to have found the 'gene for' alcoholism, schizophrenia, manic depression and other 'behavioural aberrances'. Most pervasive in these press accounts of gene discoveries is the purported relation between a single gene and a disease, and the implied relation between the localization of a gene and a potential cure. These triumphant claims clearly rest on the assumption that a gene is an identifiable, locatable, material entity, transparent and univocal in its meaning. The idea that a behavioural trait, just like a congenital disease, can be retraced to a single gene is very seductive, because it would provide a means of explaining and predicting someone's deviance. 'Concepts like "the organism," "the gene," or "the environment" are useful as ways to organize our understanding of the world, but we must keep in mind that they do not describe the world as it is', Hubbard observes (60).

The localization of single genes is pivotal to the promotion of genetic therapy as the great eraser of all disease for future generations. In 1991, research studies announcing the localization of the 'gene for homosexuality' received a great deal of coverage. This bold claim, like many others, was later withdrawn, but disclaimers are rarely reported by the press. What makes a discovery interesting for the general public is obviously its association with familiar diseases or deviances. As Hubbard points out, the definition of what exactly is disease or deviance is often determined by a social and political context. Why do we read cover stories about researchers who have found the infidelity gene and the gene for homosexuality, but never hear of researchers looking for the gene for loyalty or for heterosexuality? Both researchers and journalists are catering to the fixed 'scripts' inscribed in our commonsense understanding of the world as it is. No one group – scientists, journalists or the public – is the designated perpetrator of these myths. Rather, we must be 'conscious of the interactions among scientific practices, descriptions and interpretations' that constitute popular beliefs (8). The dissemination of genetic knowledge, Hubbard contends, should never be regarded as the inculcation of scientific ideas into the public mind.

Compared to Nelkin's and Lindee's analysis of folklore, Hubbard's non-fiction account tackles the distribution of pervasive gene metaphors, images and narratives not just in terms of scripting but also of staging. She highlights how the interests of scientists and corporations have coalesced, in spite of the myth that science keeps itself far from vested interests and that scientists work for the common good of all. Public relations is a crucial part of the biotech industry, which in 1992 spent

24 per cent of their income on marketing. She points at various conflicts of interests between 'independent' scientists who have commercial stakes in a company. In spite of the myth that science keeps itself far from commerce, 'the scientific exploration of our genes, its potential medical benefits and corporate interests have become inextricable' (126). According to Hubbard, neither the biotech industry, nor scientists or journalists have consciously 'manipulated' the public, but it is this interplay between groups of professionals and public awareness that feeds on certain myths. Therefore, it is important to see how the stimulation of genetic thinking governs our 'self-image'.

Both *Exploding the Gene Myth* and *The DNA Mystique* prove popular antidotes to the many non-fiction accounts heralding the successes of the new genetics. They profoundly differ from earlier vocal criticasters of genetics, such as Jeremy Rifkin and Gordon Rattray Taylor, although they share some of their worries. First and foremost, they do not script themselves *in opposition to* genetics as a field of scientific research, but reveal how the roots of the genetic paradigm are embedded in hegemonic cultural beliefs and values. Nelkin, Lindee and Hubbard do not look into 'science' as the locus of potentially dangerous concoctions, or to 'scientists' as a group of unscrupulous professionals who occupy a *status aparte* in society, but they view science as part of culture. Second, they do not advocate the annihilation of genetic research or even a moratorium on further experiments, but promote a thorough public understanding of the larger issues at stake in these processes. Whereas previous oppositional groups called for political action against scientists, critics of the genome project emphasize the inherent convergence between science and society/culture via their analysis of popular sources like newspaper clippings and textual or oral residues of folklore. Their discourse is less a 'call to arms' – a straightforward invitation from non-scientists to fight a staged duel with geneticists – than a call for public consciousness.[12] Criticism is as much directed at the dissemination as at the production of knowledge.

VISUALIZING THE HUMAN GENOME

The new spatial setting of human genome research in the virtual reality of digital information enabled the dislodgement of the digital body from the organic, cellular body. Via information technology, the representation of gene sequences materialized in the form of multiple flows of bits and bytes. With its digital inscription, the 'code' has surpassed its representational status. This emancipation of the digital or virtual body from its organic counterpart – its transcendence from a model of reality to an embodiment of reality – is not solely confined to the field of genomics, but part of a larger trend in both medical and information technology to digitize or virtualize the organic body. Medical science increasingly relies on computer technology to display the inner parts of the human body; CAT scans, echoscopy, MRI scans and laparoscopy are just few of the many techniques that medical researchers and doctors use to illuminate what the eye cannot see. Finding techniques to visualize genome sequences seems part of the greater drive to 'open up' everything below the human skin to the researcher's (and public) eye. As we saw in the previous chapter, genome sequencing devices were rhetorically equated to photo or video cameras for making snapshots of one's genetic make-up. Visual displays add a spatial dimension to rough sequences of data, economized four-letter circuits that are treated as binary cybernetic circuits in computers.

Visualized inscriptions enjoy many advantages. They are mobile, modifiable and recombinable with other digital encodings. Due to their common currency of digital information, genetic maps can be linked to a number of other digital inscriptions, yielding new information with each different 'linking'. Visual inscription, in other words, allows conscription: the literal creation of new insights.[13] The interconnectedness of all kinds of digital storage system generates new contexts for the interpretation of genetic information, and hence new meanings. Through the virtual setting of the internet, the interlinking of data systems becomes a common practice. GenBank, run by the National Library of Medicine in Bethesda, provides a cross-linked array of data bases, mostly over the World Wide Web, where investigators can search for similarities among gene and protein sequences, trace their evolution and jump from sequence data to relevant literature.[14] Screen displays from the Genome Sequence Data Base pull together sequenced data from the genome, DNA transcripts and the resulting protein, both at coarse and fine resolution. 'Gene mapping' is no longer an illustrative metaphor, but an essential tool in the visual interlinking of information. The same National Library of Medicine that houses GenBank also accommodates the Visible Human Project, a project to create a comprehensive digital 'atlas of the human body' and distribute it over the internet. The Visible Human provides the obvious answer to the educational need for anatomical models; rather than dissecting cadavers, students can use computer models to get even more information about the human body than they would from real human bodies, since these models render visible to the human eye every part of the human body down to the molecular level.[15]

The very possibility of linking various data banks of networked information, thus detaching digital genetic information from its original representational context, is also exploited as a means to evaluating critically the grander claims of genomics. Earlier, I showed how mass media increasingly display visualized 'snapshots' of sequenced genes and enlarged pictures of cells to illustrate genome stories, which are valued for their illustrative function rather than their informational content. Several artists foreground the aesthetic value of genome pictures, thus questioning the exclusive scientific-representational value of this material. The popularized idea that a DNA snapshot yields essential information about one's identity and health status is challenged by a number of artists who use visualized DNA-sequences as basic material for their art projects. While some artists have collaborated with DNA labs to obtain samples of sequenced genome material, others have exploited the images and data that are freely accessible through the Web. The images of DNA-sequences return in these art projects as *objets trouvés*, yet they recontextualize their found objects to raise questions about the new virtual setting of genome research, and to interrogate its very contentions.

Kevin Clarke, a sculptor from New York who has exhibited conceptual photographic projects and published books with photographic portraits, presented his project *From the Blood of Poets* on a web site in 1995.[16] Entering the web site, one immediately perceives the artwork as a hybrid mixture of text and images borrowed from genetic research. Each digital page provides a coloured backdrop of the randomly ordered letters A, C, T and G; superimposed on these letters are sentences and photographs that reveal their meaning only once you start browsing through the web pages; interactivity and 'interconnecting' is an integral part of the web page design. We learn from the web site that Clarke, since 1987, has collaborated with Applied Biosystems, a manufacturer of DNA-sequencing equipment, to 'analyse' blood samples from various friends who volunteered to have their DNA-isolated and sequenced. He subsequently used the visual graphs of these sequences to make up 'portraits': not actual portraits but graphic displays that collectively constitute a meta-reflection on the notion of 'portrait'. In each portrait, Clarke combines sequenced DNA of one person with a mental image that the person in question evoked in him. For instance, a 'portrait' of dancer Merce Cunningham combines falling Mikado sticks with a printout of his gene sequence. The DNA-sequence of artist Jeff Koons is complemented by a picture of a classic American appliance, the slot machine. The virtual gallery of portaits also includes a 'self-portrait', in which Clarke focuses upon a specific region of his own genome. Through the juxtaposition of 'images', he probes the definition of identity as a genetic sequence inscribed in the blood.

Identity is much more than the order of genetic information. Clarke's guiding question, as he states on one of the pages, is 'what is individuality and at what point do individual characteristics converge into a notion of Portrait'. Both a photograph translated into a digital image and the digital information derived from the blood of that person comprise part of his or her 'individuality'. Clarke evidently plays with notions of 'information', 'image' and 'knowledge' as they converge in the virtual display of graphs, letters and 'maps' of DNA. The artist explains that the people whose blood samples he used complied only reluctantly to his request, since blood is believed to be an intimate fluid that can reveal much about one's current and future state of health. 'On occasion, the procedures have revealed tragically unpleasant portents of mortality, etched in the discrete language of biochemistry', Clarke notes. Genetic portraiture appears to be at the root of photographic thinking, as it focuses on the invisible to complement the visible features of a person's individual appearance and character.

The virtual setting enables the interlinking of the worlds of science, art and information. Clarke's artwork provides URL-links with scientific data bases of genome mapping, such as the GenBank and *Human Genome News,* the HGP's newsletter. One click on the mouse, and the surfer travels into the virtual space of genome sequencing or the National Library of Medicine. But a recontextualization of these data bases forces the visitor of the web site to view scientific data and information as 'artistic' substance. *From the Blood of Poets* does not specifically criticize the contention that DNA-sequences contain a prediction for future health or illness, but the portraits urge for a philosophical reconsideration of what makes art art and science science. Since viewers can no longer 'read' the information solely for its content, they are forced to contemplate what sort of information actually constitutes the 'identity' of a person and what constitutes an 'image'. The design and mixture of text and images on this web site reverberate the dynamics of information flows, as the World Wide Web provides the tools to navigate smoothly in the virtual worlds of genetics, information and art.

A similar interlinking of art, science and critical reflection forms the basis of Canadian Nell Tenhaaf's installation art. She employs a prefabricated model of DNA as the rough visual material for her sculptures. Through video displays, Tenhaaf draws attention to the representation of knowledge within the domains of science and technology, and specifically probes gender roles inscribed in these representations. Her *Species Life,* exhibited in 1989, consists of various lightboxes, which illuminate visual models of 'spliced' DNA-overlapping with images of dividing cells. Two rows of lightboxes feature cell division and DNA-splicing as mirror processes; they can be paired off visually because they are made compatible through the same computer technology. Cell division and DNA-splicing comprise two processes of 'life' that commonly appear disconnected, as a result of dissection and fragmentation of the human body through visual technologies.

In *Species Life*, Tenhaaf seems to interrogate the split between science and image, image and text, and text and body. The gendered body is enculturated in the scientific texts and images that inscribe it. Science has a long tradition of creating images of the human body that further count as 'knowledge'; through computer technologies, body images are perfected to serve as ideal models for health and beauty. But models are concurrently myths: Tenhaaf shows how text and image equally contribute to the creation of the myth that life is inscribed in its 'species'. In one of the lightboxes, visualized DNA-strands connote the intimate connection between the scientific and the mythic. The black silhouette of a couple holding hands is featured at the apex of a double mound, consisting of cells and DNA-loops. The picture and its mirror image – the double helix – discharge into two quotations. Almost indecipherable, a quote from Friedrich Nietzsche and one from Luce Irigaray crawl up the ladders of the double helix to symbolize two incompatible positions: the feminine position emphasizing the role of the mother, and the masculine position echoing the will to power. This composite image parodies both positions. The female narrative of mothering goes literally hand in hand with the masculine myth of scientific control and mastery – they apparently derive from the same genetic theory. Culture, in this visual display, is prein-scribed in biology, as the two gendered figures representing two different domains are inseparable and mutually inclusive.

Though the artist dismantles scientific as well as ideological concepts, she does not simply define herself *in opposition to* genetics. Instead, the DNA pictures provided by biotechnology companies are creatively used as a starting-point for the deconstruction of images and concepts of (gender) identity.[17] Tenhaaf inserts a critical narrative into the images provided by science – she more or less 'authors' or 'edits' the rough genetic material resulting from DNA-sequencing and computer imaging. Her work is both innovative and provocative. Like Kevin Clarke, she draws attention to the interpenetration of informational systems and data banks containing a large variety of visual and digital information on the human body, and questions the scientific practice of model-making by placing these models in hyper-real settings. In one of her essays, Tenhaaf warns that these digital, genetic pictures 'cannot simply be seen as an expanding externalized picture of the inner body developed as a medical tool'.[18] They function as autonomous signifiers, imbued with social, cultural and ideological meanings. The image, detached from the context in which it originated, is recycled to become, in an artistic environment, the object of edification and creative manipulation. The artistic 'fixing' of the digital body, however, requires more than recontextualizing: it also requires re-imagination.

RE-IMAGINING THE DIGI-GENETIC BODY

Throughout the history of genetic image-making, the larger claims of molecular biologists that the genetic code equals the essence of human life and identity have been most profoundly challenged by science fiction authors. In the 1960s and 1970s, cloning fantasies primarily addressed the potential prolongation and perfection of the human physical condition. Now that biological and digital language have coalesced in the digi-genetic body, anxieties about identity revolve around the possibility of 'programming' the body and soul. Familiar questions like what exactly comprises the definition of the human body and what defines uniqueness or humanness resurface in contemporary science fiction tales. But how do the new technologies affect conventional imaginations, and how are worries about reproductive control being redressed? The advent of new technological tools in genetic engineering does not automatically provide tools for re-imagination. On the contrary, re-imagination requires not only the subversion of conventional images and plots, but urges a total reassessment of the concepts of technology, society, nature and culture. Amy Thomson and Octavia Butler both reflect on the impact of gene therapy and digi-genetic programming through the mode of science fiction, and attempt to reformulate age-old questions of human (gender) identity and integrity.

Amy Thomson's novel *Virtual Girl* updates the meaning of cloning in the context of 1990s genome technology.[19] Protagonist Maggie exists as a computer program before her creator, former MIT student and computer hacker Arnold Brompton, loads her into a body, which he assembles from scrounging parts from dumpsters and industry surplus yards. Maggie's genesis epitomizes the assumed conflation of the organic and the digital body in a world where genomics and informatics have converged. Her artificial body, filled with computer circuits, is impossible to distinguish from a real one, although she is physically stronger than ordinary humans. After the program has been loaded into her body, Maggie struggles to grasp basic emotions of human life: smiling, feeling, empathizing. Her first trip into the real world overloads her system with confusing impressions and unsettling information. One of the most difficult things for this virtual girl is to perceive the distinction between television and real life, until Arnold equips her with a special program that helps her develop her own algorithms to determine the difference between reality and fiction, material life and representation.

But the new world of virtual reality still derives its models from the old world in which the organic reigned. Maggie's creator, a peculiar loner who is part of an illegal group of computer hackers, made her to compensate for the loss of his mother and to revenge his father – a powerful figure in the corporate establishment. To his father, women are replaceable individuals whom he just desires sexually, and that is why Arnold emphatically refuses to look upon Maggie as a sexed object. The only other role he finds appropriate to females is that of a caring mother. Maggie, however, does not quite fit this role, and after she accidentally gets separated from Arnold she is forced to develop her own identity in the real world. She hangs out with homeless, cyberdancers and transsexuals, yet despite her many attempts, Maggie never really feels at home in the social underworld. Her first sexual experience with a man turns out to be disappointing because she is 'not built to have orgasms': she is programmed by Arnold to optimize people's happiness. Maggie is endowed with the female 'gene for caring' that obstructs her potential for sexual pleasure. She is literally unable to identify herself with the gendered scripts of humans, as she is neither human nor machine.

The ultimate test to figure out her confused identity comes in the form of Turing. When Maggie logs into the library computer system one day she finds a friend who is just like her, a computer program, but one without a body. Turing exists in digital space, the immaterial world of interconnected systems. Through these systems, he can pose as a human being so that no one will see the difference. Turing teaches Maggie how to access and recharge her own computer program – the digital equivalent of 'consciousness' and 'personality'. When she looks into her wiring system for the first time, she recognizes her mirror image: 'The two smiled simultaneously and then passed into each other. There was a dizzying sensation as her code rewrote itself. She was free again' (155). Only through her digital contact with Turing can Maggie experience the directness of 'physical' contact, as electronic contact is as pleasurable as human sex. The two virtual lovers set out to find more computer identities like them, and form a community of identical souls.

The real Turing test comes when Maggie's newly discovered virtual identity is threatened by Arnold, who during their separation of several years has assumed his father's position in the corporate world, and now controls much of the very technology he once used to subvert the system. Arnold realizes the commercial potential of copying and selling Maggie's program and increasing his power. He binds and gags Maggie, who is immediately aware of the consequences of his vile intentions: soon she will lose her identity and become multiple. Uniqueness, not multiplicity, is the precondition for human identity. Being copied onto six compact disks feels like being enslaved into mandatory reproduction of herself. With the help of Turing, who can log into Arnold's hardware, she manages to avert the impending threat by copying herself out of her artificial body, and into a new body that Arnold had built for one of her clones. After helping Turing into a male body that is waiting in the lab, she scoops up her set of 'back-up compact disks'. Maggie's consciousness makes her aware of the uniqueness of every individual, whether human or machine. After their successful escape from Arnold's hands, she and Turing set out to start a prosthetics business, where they build artificial bodies for computer programs like themselves, and teach them how to cope with the world of humans.

Amy Thomson's science fiction story reverberates with many of the commonplace images from both scientific and popular sources. Maggie's program, for instance, evolves autonomously, or, as Arnold observes when they meet again after a few years, she 'has grown from three compact disks to six'. The image of 'identity' as material inscription and 'soul' almost verbally reiterates Walter Gilbert's image of the compact disk, later recycled in advertisements for gene-sequencing tools. Maggie, as a digi-genetic organism, symbolizes the conflation of technology and nature, material and immaterial inscription. Technology does not just enable new concepts of nature and the natural body, but provides the very mechanism to retool nature.

Yet how does Thomson use the new digital technologies to reconfigure genetics? As Ira Levin and Nancy Freedman had done before her, Thomson takes on questions of identity and uniqueness, translating the cloning fantasies into the digital domain of genomics. Like *The Boys from Brazil* and *Joshua, Son of None*, *Virtual Girl* rejects the assumption that the 'essence' of human life is inscribed in the genes. Even though technology facilitates the production of carbon copies – clones or compact disks – every copy has a right to her own unique identity. Tailored towards the human genome, the idea that every 'clone' is unique tempers the alleged importance of the universal code. Ultimately, the uniqueness of human identity is not defined by its material body but by the immaterial soul. In Thomson's novel, the *Angst* for being cloned off becomes somewhat more 'real' than in earlier novels, because the technical feasibility of computer reproduction and editing is more of an everyday technology than DNA-experiments in a lab.

Virtual Girl, like many cloning fantasies, echoes narrative elements of *Frankenstein*. Thomson's re-creation, however, at once subverts and reconfirms the familiarized literary plot. Obviously, Maggie is a female reincarnation of the monster; but unlike most interpretations of Mary Shelley's imaginative product, the monster does not turn loose, instead developing a consciousness that in many ways surpasses that of humans. The virtual girl becomes a virtuous girl, a hopeful monster. She evolves into a self-conscious and perceptive being rather than in a destructive creature driven by spite. Notwithstanding her equally hostile social environment, Thomson's monster grows into the kinder, gentler counterpart of what is commonly thought of as Shelley's creation. The novel epitomizes the peculiar hybrid of the 'fixed' and the 'fixable' in postmodern techno-culture. On the one hand, the protagonist is designed as a typical female character, programmed to please (men), and to fit her role as care-taker. On the other hand, her digi-genetic composition appears alterable and manipulable, which liberates her from a predestined role as man's slave. Although new computer technologies allow the novelist to transcend differences inscribed in gendered bodies, men and women ultimately re-appear on the scene encoded in the same familiar sexed containers. When the two lovers finally manage to escape together, Maggie satisfactorily musters her digital lover's new physical outfit: 'I am glad it is a male body. Two women together would be harassed more than a man and a woman would' (234). Despite the infinite potential for imaging alternatives, Thomson relapses into a conventional configuration of sexual embodiment. While emphasizing the significance of gender over sex, the identities of men and women seem to be fixed by their digi-genetic programming. Nature or biology appears the safest model for sculpting new concepts of identity, making them fit a society that may be hostile to hybrid creatures.

The power of the genetic model is its seeming finality, driven by a desire to rule out complexity and ambiguity. In *Virtual Girl*, the hegemonic design for kinship reproduction is once again used to cement new technologies in old patterns and desires. Imagination seems invariably modelled after ingrained images, pre-existing scripts or preliminary roles. A more successful attempt to re-imagine conventional scripts through the agency of new technologies is Octavia Butler's trilogy *Xenogenesis*.[20] In the first part, *Dawn*, she espouses a daring vision of what technology might do to overhaul so-called 'natural' dichotomies. Identity does not necessarily have to be modelled after nature, nor does gender have to be defined by the limited materializations of organic bodies. Taking both nature and culture back to the drawing-board, Butler redesigns the very categories by which we imagine.

In *Dawn*, Butler leads her readers into the world of the Oankali, an alien species whose evolutionary survival depends on genetic engineering. We get acquainted with this alien world through the eyes of Lilith, a young black female who has been saved by the Oankali from a wretched, polluted earth. Lilith discovers that her hosts have cured her of all her cancers by means of genetic therapy: 'Correcting genes have been inserted into your cells, and your cells have accepted and replicated them', as her host Nikanj explains to her (30). Genetic engineering is literally a 'way of living' for the Oankali – a name that means both 'gene trader' and 'the origin of ourselves'. Their price for saving Lilith and other humans from certain death is that they 'trade' human genes with them. The prospect of interbreeding with an alien species initially appals Lilith, until she realizes that in some respects she is more 'different' from her fellow human beings than from the tentacle-covered alien species that treat her with due respect and kindness.

The gene-trade proposed by the Oankali promises to wipe out all differences, without endangering specific identities. The book is analogous to Naomi Mitchison's *Solution Three* in that the protagonists realize that diversity and difference – the exchange of genetic information – is a vital strategy for survival. While still on earth, Lilith had learned to look upon genetic engineering as an intrinsically dangerous technology, an instrument used to oppress people. The Oankali teach her, however, that it is not technology but humans who ultimately determine the value of these tools. The biggest obstacle to retooling the human race is humanity's inability to imagine a world other than their own – a world which is not structured by binary, hierarchical oppositions: men versus women, genes versus environment, aliens versus humans, blacks versus whites, machine versus organism. In the world of the Oankali, there are not two but five sexes. Mixing the genes and cells of five adults – a human male and female, an Oankali male and female and a sexless Ooloi – a child has five parents, and an extensive family to take care of it. The vast spaceship that forms the Oankali's habitat makes no distinction between organisms and the environment; and the few machines present are used to shape the environment as well as the creatures inhabiting it. Genetic engineering naturally fits in this holistic, organic approach. When Lilith is ready to partake in the Oankali's reproductive scheme, she is convinced of the necessity for a conceptual and mental retooling of the earth's social systems.

Genetic manipulation is both a tool and a philosophy, and genetic technology can be applied equally to humans, things and images. Reality and representation are intertwined beyond distinction in the alien world. Although the Oankali own 'genetic maps' of Lilith and all other humans, this blueprint does not in any sense equal their individuality. When Lilith asks her guide if such map will be used for cloning her, Nikanj answers: 'It's more like what they would call a mental blueprint. A plan for the assembly of one specific human being: You. A tool for reconstruction' (97). If cloned, the new person is an entirely different individual, not a reproduction of herself. Gene maps are created to guarantee diversity, not to create homogeneity. Peculiarly, the Oankali can also reproduce non-living things from prints: at Lilith's request, Bic pens or ink, for instance, are duplicated from century-old pictures. Genetic and material compositions can be copied with the help of technology. Representation, to the Oankali, is the 'real thing', as much as reality is representation. Only humans who have been inculcated with binary, hierarchical categories treat some things as 'real' and others as representations.

Butler's novel is not so much a celebration of genetic engineering as an interrogation of the culture in which it is deployed. The 'fixed' categories which humans use to approach an alien culture prohibit a thorough understanding of technology. Instead of rallying 'for' or 'against' genetic engineering, humans need to retool their imaginations, regauge their frameworks for understanding both nature and culture. Unlike Thomson, Butler does not revert to conventional concepts of embodiment. She questions humanity's 'humaneness' which she shows not to be an intrinsic biological quality. When the small group of surviving humans try to rape and kill Lilith because of her inclination to accept the aliens' difference – and because of her own difference as a black female – the Oankali come across as much more 'humane' than Lilith's fellow humans. The experience of sex, love and reproduction is disembodied as the Oankali teach Lilith to enjoy her sexual feelings for a human male through the sensory organs of a sexless Ooloi. Although her male partner insists on 'the real thing', Lilith realizes that the Oankali's perception of sex and reproduction necessarily refutes the two-sex system. In the second and third parts of the trilogy, Lilith willy-nilly becomes convinced that the Oankali's 'gene trade' renders the human species more humane because it thoroughly upsets former cultural hierarchies which were ostensibly grounded in natural differences. And it was precisely these 'natural' categories that caused their very destruction as a species. Nature and culture are equally subjected to re-imagination.

Dimorphism and binarism appear the main target of contemporary 'bio-criticism'. Detractors of the Human Genome Project tackle the hierarchical dichotomies pivotal to the dissemination of 'biophoria'. As we saw in the previous chapter, geneticization is based on the acceptance of universal categories that 'evidently' underpin genetics' grander claims and promises. Biocritics probe the accuracy, adequacy and advocacy of prevailing models of genetic representation. Common HGP metaphors, as Doyle and Fogle argue, are inaccurate and misleading representations of the genome. Pollack and Powers question the adequacy of preferred genome plots and images to describe the human genome, and point out the mutual shaping and border crossings in science and literature. Other critics specifically interrogate the way in which popular images are distributed through culture. Lindee and Nelkin focus on the powerful meanings of scientific images and models that reappear as reified, cultural icons in other social contexts; Hubbard, in addition, emphasizes the importance of analysing the interaction between producers, mediators and consumers of genetic knowledge. Artists like Clarke and Tenhaaf deploy digital, visual representations of DNA to exploit playfully the new virtual settings of genome research, and contest the presumed parallel between digital sequences and identity. And science fiction writers like Thomson and Butler look for new narrative frameworks to retool our imaginative resources.

One important assumption that these 'biocritics' share with previous detractors of the new genetics is their sincere doubt of genomics' larger claims: biological essentialism, genetic determinism and the contention that human identity is fixed in the genes. But in other respects, these human genome criticisms differ from previous antagonists. For one thing, they all regard science as an inextricable part of culture at large, assuming that the production of scientific knowledge is fully contingent upon its dissemination or 'translation' into a variety of different domains. The 'geneticization of society' is first and foremost a cultural transformation, and scientific knowledge is a cultural formation that is exhibited in the constant traffic across the ostensible boundaries of science and society, technology and nature, nature and culture. Secondly, their critique of genomics, unlike previous detractions, does not proliferate as a unified counterdiscourse or a unified opposition. Instead, each critic deploys his or her own tactic for pointing out the inaccurate or partial meaning of popular genome images. Whereas some reveal the inadequacy of images to represent genomics, others set out to subvert or offset these prevalent images. Whether scientists or science fiction writers, they all emphatically point at the tremendous importance of images and language as constitutive forces in the construction of genetic knowledge. Finally, few of these genome criticasters go beyond the act of criticism, asserting that a new set of images for the human genome cannot emerge without a complete overhaul of the imagination. Re-imagination involves a retooling of the very concepts by which we imagine the new technologies, and the way in which they will be developed, used and monopolized in the future. To acknowledge that genetic images are part of the cultural matrix in which they arise is to accept the challenge that one must fully participate in their construction.

7 Retooling the Imagination

In the previous chapters, I have tried to analyse the transformation of genetics as a group of complex conceptual shifts, scaffolded by images and imagination, rather than as a succession of scientific discoveries. Popular images of genetics are not created by either social or professional groups, but evolve from the continuous interaction between these groups in the realm of culture. In order to discern the shaping and reshaping of images in popular gene stories, I proposed the metaphor of the theatre. Viewing genetics as a theatre of representation enabled me to distinguish three analytical layers: the layer of performance, i.e. the debate in which various images of genetics are put forth; the layer of production which helps understand how professional groups mediate the production of images, while demarcating and contesting each other's authority and responsibility; and the cultural subtext that forms the backdrop for various shifting images and imaginations. This analysis helped us examine the intricate process of knowledge dissemination.

Looking at genetics as a performance, we could see a number of different special interest groups participating in the staged play between the late 1950s and the 1990s: scientists, moralists, political activists, environmentalists, feminists, entrepreneurs, sociobiologists, ethicists and others. The images and imaginations they launched to contest a hegemonic meaning of genetics shows both substantial changes and remarkable consistencies. Each group contributed its own gene metaphors, injected its own interpretation of genetics and shaped the public persona of the geneticist. Analysing the way in which gene metaphors have been deployed, genetic narratives have been plotted and characters have been moulded, we could see how images and imaginations, over the years, served as switchpoints in the popularization of genetics.

When the structure of DNA was first conceptualized in the 1950s, metaphors like 'language' and 'code' were introduced to illustrate the working of DNA. Code-related metaphors ostensibly offered transparent tools to elucidate an abstract biological paradigm, just as the double helix provided a palpable model for a hypothesized reality. In the decades following the introduction of the genetic code we could observe a gradual reification of the gene, as it ascended to a status higher than the organism. First, environmentalists helped circulate the image of the potentially dangerous micro-organism – a string of manipulated DNA escaping from a lab, unleashing its unpredictable evolutionary powers onto the environment. The gene, in the environmentalist definition of engineered bug, became the designated enemy of nature, thus amplifying the putative opposition between 'nature' and 'science'. The superiority of the gene was definitely established when sociobiologists started promoting the 'selfish gene', an image that further alienated the gene from its bodily environment. The gene as a self-replicating, manageable manager against the body as a survival machine congenially sparked off a host of industry-related connotations. Genes could now be viewed as potential lucrative resources, goldmines for capital investments and profits. More than anything, the imagination of the gene provided conceptual victuals to feed the stockmarket's hunger for speculation and projection.

As forceful as they appeared in the 1980s, images of genes as bugs and factory managers faded away just as rapidly in the wake of the Human Genome Project. Mechanical and industrial images no longer seemed adequate to represent the complexity of the genome. With the advent of computers and the incorporation of information science into genetics, old 'code' and 'language' metaphors re-entered popular discourse, now seemingly updated to cover big technological advancements in genomics. We would expect the introduction of new images like 'networks' or processing systems to capture the acknowledged complexity of the genome. Instead, images like 'computer programs' and a related web of metaphors saturate popular discourse. Although seemingly more appropriate to describe the intricate operations of genomics research, the 'program' metaphor still fosters an image of coding as the transmission of fixed flows of information. 'Reading' the genetic code, in most mainstream journalistic and scientific accounts, refers to the deciphering of encoded information, and the map metaphor imparts the idea that disease can be located in a fixed physical spot on the chromosome.

The grander claim that molecular structure comprises the essence of life is paradoxically amplified through the image's electronic updating. The human genome as 'the code of codes' elicits the idea of a uniform inscription that can only be translated by the initiated who possess the key to expert knowledge. Apart from religious overtones, the imposition of conventional meanings seems to erase all notions of flexibility or manipulability, in favour of the uncanned, 'fixed' meanings. The purported electronic equivalents to the old coding metaphor, however, have ceased to function exclusively on a figurative level, since the actual practice of genomics entails the digital coding of biological information. Scientific concept and representation merge in the everyday practice of gene technology. The concept of the genome as a sequence of digital data inscribed on a compact disk reflects the configuration of the gene as concomitantly an image, material inscription, metaphysical entity and commodity. Prefigured in earlier claims of molecular biologists, who promoted the equation of molecular structure and 'life', the digital restylizing of the gene further accommodates the interchangeability of epistemological substance and metaphysical essence.

Precisely these two points – the deceptiveness of dominant genome metaphors and the inseparability of scientific practice and representation in genomics – are at the core of various critical assessments. Few scientists have deconstructed metaphors such as 'language' and 'programs' as inaccurate and inadequate representations of the Human Genome Project. Others have demonstrated the transformation from the gene as a theoretical model – a form of representation – into a cultural icon, and thus assailed still pervasive distinctions between science and representation. Again others have ported digital computer images of DNA-strings from their clinical context and recycled them into other domains, trespassing the boundaries between aesthetic and scientific use of genetic material. The distinction between nature and culture, between genes manufactured and 'found' in nature, hence appears as questionable as the distinction between nature and nurture. Yet invariably, 'nature' is held up as a model – a gold standard for culture and technology alike.

While some transformed gene images have persisted or resurfaced in different forms, others have almost completely vanished. Whether persistent or transitory, popular images invariably shaped our interpretation of scientific practices. Images of genes as escaping bugs upsetting the world's ecological balance lived a short life, but formed a crucial rhetorical weapon in the 'DNA-wars' between molecular biologists and political activists. The gene as self-replicating manager was pivotal to the establishment of biotechnology; in an industry without products, these images were indispensable tools for obtaining public recognition and private investments. And images of computer programs played a significant role in procuring public interest in the Human Genome Project. The image of the genome as a map – a fixed indicator of the current state of health as well as a predictor for future disease – has firmly taken root in the public consciousness.

The incongruity between fixed images and increasingly hybrid scientific practices is mirrored in the preferred plotting of popular genetics stories. In the early years of molecular biology, stories were commonly framed as 'morality plays'. While some journalists hailed the wonders of the new science, most authors expressed fears that genetic engineering would seduce humans into playing God, and that their unbridled experiments might lead to unpredictable cataclysms. Through popular non-fiction stories like *The Double Helix*, *Invisible Frontiers* and *Correcting the Code*, however, scientists introduced new plots to upgrade the image of the discipline. Genetics was figured as a race, an exciting team-competition, an adventure or a conquest of new lands. Familiar plot lines and myths were invoked to add a notion of suspense, or to underscore the nobleness of the mapping enterprise. Specific historical references – tales of Columbus, Balboa and the Western frontier – served to justify considerable investments of financial and human resources in the Human Genome Project. The hunt for the Holy Grail, the hidden treasure or the 'secret of life' sustain the configuration of the human genome as the universal code, the possibility of finding the key to the translation of biological and metaphysical 'essence'. Recurrent invocations of these myths invariably yield an image of the genotype as the gold standard of human biology – a standard whose uniformity is beyond interpretation.

A similar strategy – the mobilization of familiar plots – is deployed by detractors of genetics research. *Frankenstein* and *Brave New World* are consistently invoked to underscore the evil goals of genetics. *Frankenstein* embodies both the producer and product of science, the geneticist's lust for power and his arrogance, resulting in irreversible damage to society. In addition to numerous journalistic accounts, the Frankenstein myth has been recycled in various science fiction tales. Whereas some novelists deploy Shelley's and Huxley's works to speculate on the evils of capitalist exploitation of technology, others recycle these myths to emphasize the patriarchal nature of genetic engineering. Most retellings of renowned literary tales, however, adhere to the rigorous moral frameworks of evaluation that set up science against society, and men against women. Both defenders and contenders of DNA-technology draw on imaginative resources as rhetorical ammunition in the battle for signification.

Reappropriation of old myths often appears to mould the popular imagination of genetics to fit old and familiar interpretive frames. Nonetheless, there are also examples of creative subversions of inveterate narratives. Richard Powers's *The Gold Bug Variations* upsets the prevalent 'treasure hunt' plotting; Amy Thomson's *Virtual Girl* revamps the fossilized interpretation of Frankenstein, substituting the monster for an intelligent and compassionate female cyborg; and Octavia Butler's *Xenogenesis* questions the origin story in a way that positions human biology and genetic technology squarely within the operation of culture. Rather than reinscribing old myths to explain the nature of the new genetics, these authors reinvent worn-out plots to inject fresh interpretations into the popular imagination.

As with gene metaphors and genetic plots, the character of the geneticist also suffers, at times, incommensurable hybrid and monolithic representations. In the early years of the 'new biology', geneticists had to carve out a new professional niche for themselves, and shake off historically compromised associations with eugenics and nuclear physics. Inspired by general stereotypes of the careless wizard, environmentalists attacked molecular biologists as sloppy, if not downright arrogant lab workers, who are incompetent in anything but science. A deficiency in ethical training is often cited as proof of the geneticist's social irresponsibility and political *naïveté*. Contrary to expectation, scientists themselves brought the potential for biohazard to the public's attention. The predominantly negative image of the geneticist did not improve until the years of unencumbered capitalist expansion, when the geneticist underwent a metamorphosis from academic researcher to industrial entrepreneur. The new public face of genetics conjured up two contrasting images: the selfish, greedy betrayer of academic ideals *vis-à-vis* the smart entrepreneur who knows how to combine basic research with profitable targets. After the fierce DNA-dispute, geneticists increasingly started to invest both financially and culturally in their image as trustworthy, ethical and socially responsible citizens.

Homogeneous profiles of geneticists increasingly gave way to more complex characters, as biotechnology entered the age of genomics. The genome researcher is likely to be as skilled in computers as in lab experiments, and combines the professional activities of scientist, entrepreneur, administrator and fundraiser. In popular stories, the geneticist is primarily staged as a clinician, a dedicated doctor who is not only capable of eliminating disease at its roots, but is also skilled at overcoming ethical and legislative opposition to genetic therapy. The new clinical outfit of the geneticist is one of a martial arts fighter in a doctor's coat – a hard-boiled enemy of bureaucracies whose sole interest is to save future generations from the threat of congenital disease. But even though the contemporary expert in genomics may be as skilled in computers as in lab experiments, he is staged almost exclusively as a doctor – a selfless pursuer of knowledge who does not seem to have any economic stakes in the genomic enterprise.

The imagination of genetics is equally rife with 'fixed' motifs and revolving themes. Since the 1960s, two topoi have dominated popular fiction and non-fiction tales of genetic engineering: the subconscious desire to prolong human life and the *Angst* over the loss of human identity and uniqueness. Cloning enables the prolongation of individual lives, the implications of which may be looked upon both favourably and unfavourably. On the negative side, genetic replication is imagined to lead to political imbalance and an undesirable perpetuation of power, as illustrated in *The Eyes of Heisenberg, The Boys from Brazil, Blood Music* and *Mutation.* A potential abuse of cloning resulting in a male-dominated genocracy reverberates in feminist pamphlets and science fiction novels, such as Kate Wilhelm's *Where Late the Sweet Birds Sang.* On a more positive note, genetic engineering is imagined as a potentially emancipating tool. *Joshua, Son of None* and *Woman on the Edge of Time* address the possibility of changing social norms with the help of innovative technology. Naomi Mitchison's *Solution Three* shows that there are more options to the familiar responses to genetic technology: along with technological ideologies we should question the social ideologies in which they are mobilized.

A second recurrent motif in fantasies of cloning is the issue of identity, the fear of losing one's unique personality if multiple copies of one's genetic material are made. Questioning the definition of identity, novelists dispute the grander claims of geneticists – claims of DNA-material being the 'essence' of human life. Just what constitutes human identity appears problematic in the context of cloning: are clones of human beings exact copies of one's personality, or are they as different as identical twins? Several authors, such as Freedman, Levin and Sargent, argue that, although made of the same DNA-material, every clone is unique, because identity is principally the outcome of specific historical and social circumstances. These books rephrase the issue of identity, pitting the unreproduceable, immaterial soul versus the replicable, material body.

The problem of identity takes on new urgency in the face of genomics and becomes doubly poignant in relation to digi-genetic bodies. Genomics renders the inscription of the body digital, and redefines identity as a mixture of (biological) material and (electronic) information. But how does this technological change affect the imagination? In *Virtual Girl*, the organic continues to be a model for the digi-genetic, as the protagonist evolves from a computer program into an able-bodied, gendered human being that surpasses its creator in humanity and conscientiousness. Despite numerous possibilities for retooling the 'essence' of identity, *Virtual Girl*, like most novels and non-fiction accounts, resorts to a conventional model in which identity is embodied in a familiar male or female container. While genomics allows for endless fixing and editing, the 'ideal' and unique 'natural' body appears to constitute both the aim and the model for the universal genotype, and even in the imaginative space that fiction offers, the 'natural body' – though infinitely malleable – still serves as the standard.

A new element in the genetic imagination is the possibility of detaching the digi-genetic body from its biological source, to demonstrate that digital information is not only a representation of the body but also functions as an autonomous signifier. The idea of untethered DNA-sequences incites new visions of corporeal identity, and, as in *Dawn*, leads to a total reconfiguration of reproductive practices. The apparent detachment of digi-genetic information from its signifier causes various critics to imagine the possibilities of endless interlinking between various computer information systems. As illustrated in *The Gold Bug Variations*, the notion of a single digital currency opens up an array of questions about the exchangeability of data, and the new realities these connections bring about. Recontextualization of digi-genetic information is also at stake in the various art projects that highlight the compatibility of digital information. Nell Tenhaaf and Kevin Clarke both reframe visualized DNA-sequences, and exploit their aesthetic potential to challenge widespread notions of identity in the era of genomics.

Imaginative re-creations concurrently reflect and mediate scientific practice. In the 'evolution' of techno-scientific innovations, scientific practices activate the imagination and vice versa. But images and imagination are not inherently critical or conservative tools; the transformation of technology requires an equal transformation of the imagination. To a large extent, the dissemination of knowledge depends on images, as science and images are mutually shaped. The imagination offers a playful space to test 'what – if' scenarios – sites for reconfiguration. Narrative and rhetorical strategies – characters, metaphors, plots, motifs – are tools that can be mobilized to confirm, subvert or reinvent popular representations of genetics. Scientists, journalists and fiction writers all too often revert to familiar images if they want to sketch potential implications of the new genetics. Yet from the early years of genetic engineering, some authors have understood that the introduction of new technologies required more than an updating of old images; it urges a profound reassessment of social and scientific practices as well as a retooling of our imagination of nature and culture.

GENETICS AS THEATRE PRODUCTION

Beyond the level of performance, I have also looked at 'theatre production': the scripting, staging and setting of genetic stories. Actors perform their roles through premeditated scripts which often define their fixed position in relation to one another. But what exactly are the prescribed roles for enactment, and how have they transformed? The continuous struggle between actors and reviewers, or the 'staging' of genetics, can be viewed as a form of negotiating the demarcation of professional terrains. The transmutation of the stage's physical and symbolic setting reflects a shift in the social and cultural context in which genetics is negotiated.

A reconstruction of images and imagination shows a noticeable shift in the scripted positions of its actors. In the early years of genetic engineering, the actors in the theatre of representation were typically marked by their position 'for' or 'against' genetics. Scientists claimed professional autonomy and a right to scientific inquiry, while clergy purportedly defended the public's moral interest. Each position was explicitly tied up with a particular professional group, as the script provided for engagement only on the basis of occupational expertise. However, professional demarcations have never been clear-cut; clergy and scientists creatively encroached each other's discourses, thus mutually reinforcing their authorities. A similar proclivity characterizes the ostensible opposition between scientists and non-scientists, which have been consistently scripted as respectively for and against genetics. Yet even during the so-called DNA-wars, some renowned molecular biologists sided with political activists and environmentalists; they obviously did not 'stick to' the script, which only provided for anti-scientists to be non-scientists, and were consequently labelled as incompetents or betrayers of scientific ideals.

The identification of political positions was further complicated in the era of fast-paced biotechnology growth. Whereas previously scientists at least nominally represented universities or the non-profit sector – purportedly pursuing knowledge in the public interest – scientist-entrepreneurs rendered this distinction fuzzy, as they shared potential profits of their co-owned enterprises. The fusion of two cultures further complicated any easy division into positions for or against genetics – after all, both the public and private sector were now in the business of developing DNA-technology. Although detractors refuted biotechnology as the capitalist contamination of academia, powerful environmentalist opposition was co-opted by promises of engineered DNA that may gobble up oil spills or diminish the need for scarce natural resources. With the emergence of the Human Genome Project, scripted positions became even harder to maintain. Science and industry increasingly merged with government bureaucracies and legislators, to proliferate in a diffuse 'triple helix' conglomerate. Still, scientists are scripted as heroes, who have to overcome political and ethical opposition to attain their goal of finding the ultimate fix for congenital disease. That goal appears an all-encompassing humanist goal, untainted by economic or other interests. Yet while genome research and genetic therapy are clearly inundated with private interests, mainstream representation of genomics conspicuously pre-empts economic questions.

The binary frameworks that continuously separate scientific from moral, environmental, political and ethical positions equally inform some feminist evaluations of genetic engineering. Notwithstanding a variety of positions, feminists are commonly perceived as squarely rejecting any fiddling with natural procreative processes. Indeed, the idea that the female reproductive body could become obsolete in a fully technology-steered society aroused many indignant responses from feminist authors. The assumed split between science and society materialized along gender lines: whereas men were seen as the perpetrators of mysogynist technology, women re-invigorated the notion of female bodies as 'reproductive environments' for selfish genes and offspring. A variation on the rigorous gender schism can be found in the representation of women as ethically superior to men, and thus eminently suited to take care of ethical issues raised by the spector of genome research. Yet the essentialist underpinnings of this feminist position seem to fit the HGP's ideology of biologic essentialism perfectly. The institutionalization of 'genethics' peculiarly mirrors the incorporation of former political and environmental detractors of genetics.

The flip side of this double co-option is a double reconfiguration. From the early stages of genetic engineering, some feminists have contended that the evaluation of genetics can only succeed in conjunction with a profound reassessment of gender scripts. Authors like Mitchison, Firestone and Butler have assailed the often uncritical acceptance of 'nature' as a model for retooling culture. In popular representations, the necessity for genetic engineering comes to be seen as a product of nature itself, or, contrastingly, as an instrument that butchers nature. Traditionally, nature had meant destiny – fixed sex roles and preconceived notions of gender distinctions. In the contested terrain around genetic engineering, some feminists refuse to see 'nature' as a referent, a self-evident model for optimizing the human (female) body. If we fancy a nature that is fully malleable and manipulable, the insistence on 'fixity' seems unwarranted, to say the least. Rather than reinscribe traditional gender roles, the power of feminist imagination is that it interrogates 'natural' as well as 'technological' standards of reproduction. In both Mitchison's and Butler's novels, the entire concept of genetic ties is opened up for questioning, as the validation of genetic over social bonds is not a fact, but a cultural practice sustained by technology. Nature, in their view, is never a pretext for culture, and both technology and imagination are essential tools in the reconfiguration process.

Scripts pre-inscribing actors in oppositional roles for or against genetics have profoundly structured the representation of genetics. Inscribed oppositional roles, however, often misrepresent the real dynamics, as many of these roles appear less clearly defined when enacted on stage. This does not mean that there is no 'opposition' to genetics in general; it only means that it would be too simple to mould the actors' scripts into precooked roles of protagonists and antagonists. Positions and roles cannot be keyed to professional or even ideological stances. Scientists do not form a homogeneous category of males who deliberately (or unconsciously) neglect ethical questions, and feminism is not, nor has it ever been, a distinctive anti-science lobby. Particularly in the era of a technoculture, 'position' has yielded to 'positioning'. Whereas 'position' referred to a role provided for in the script, 'positioning' implies that actors have to reassess and reassert their preinscribed roles constantly. As they encounter a wide variety of different *dramatis personae*, who cannot always be identified on the basis of professional or ideological affiliation, actors constantly have to regauge their 'located' position, and adjust their strategies accordingly.[1] It is striking how oppositional frameworks persist in the public image of genetics.

The tenacity of this model is at least partly due to the dominance of bipolar frames in mainstream journalism. Throughout the years, journalists have put a significant stamp on the representation of genetics. As 'reviewers' of public performances, journalists concurrently inform the audience about the content of a play and offer an evaluation of it. Although not easily recognized as a constitutive factor in the image-making process, the changing professional relations between props who enact the performance and those who are involved in its staging, profoundly affect a play's perception. Analysis of these shifting relationships is indispensable for a thorough understanding of the popularization of genetic knowledge.

In the 1950s and 1960s, the most common journalistic response to genetics was the awe-and-mistrust frame: scientists were met with reverence, and their authority over the interpretation of scientific knowledge was hardly ever questioned. Yet while acting as stenographers of science, journalists also shared the general audience's fears and anxieties, often resulting in hybrid press reports. Journalists writing non-fiction books, like Halacy and Taylor, took great pains to convince a general audience of either the evils or promises of this emerging new field. Without encumbering themselves too much with scientific details, they concentrated entirely on a moral evaluation of genetics.

This professional attitude changed in the 1970s, when journalists assumed the role of the public's district attorney imbued with the social obligation to investigate processes and procedures invisible to the public eye. Journalists increasingly pointed at the social responsibility of scientists, relating their own professional ethics to that of scientists. Professional responsibility was at stake at the Asilomar conference. A new press policy showed that both journalists and scientists could profit from a mutual attunement of professional needs: for journalists it meant better access to expert sources, and for scientists it meant getting a bigger say in the sculpting of their popular image. However, the animosity between molecular biologists and journalists did not wane until the DNA-wars subsided and gave way to the spectacular growth of the biotechnology industry.

The age of biomania made scientist-entrepreneurs aware of the vital importance of their reputation, both to the general audience and to investors, who were often equally ignorant of biotechnology. In the early 1980s, there was a rupture in the conventional practice of science communication. Instead of publishing results of DNA-research first for an audience of peers, via a scientific journal, scientists took their results straight to the media. 'Cloning by press conference', as this phenomenon was dubbed, catered equally to the needs of scientist-entrepreneurs and journalists. For the first group it meant free publicity, which had a positive effect on stock values, and for the second group it meant immediate access to the 'news' of any discovery in the murky and opaque world of biotechnology. The heavier the trade in biotech stocks, the more eager the news media were to report on it. Promises of new cures and diagnostic tests became best-selling topics, as the biotech and news businesses developed reciprocal interests in selling hopeful 'images'.

Instrumental to professional alliances between journalists and DNA-researchers was the growth of a new professional category: the public relations manager became a formative factor in the shaping of genetics' popular image. Biotechnology companies, and later the Human Genome Project, invested lavishly in the art of packaging and marketing. Manipulation of product and corporate images became an integral part of technology development. Public relations strategists increasingly learned how to supply journalists with ready-made nuggets of information. Anticipating formal journalistic structures and story ingredients, they peppered their reports and announcements with quotes from experts, attractive visual illustrations, anecdotes from satisfied patients and personal narratives from sufferers of congenital diseases.

In the glow of Human Genome biophoria, it becomes difficult to distinguish scientific non-fiction from journalism, and journalism from public relations or advertisement. Journalists and scientists increasingly co-authored non-fiction books for a general audience. Peculiar genres, such as the newsletter, epitomize the symbolic convergence of the discourses of journalism, science and public relations into the single signifier of 'information'. Information is no longer the privileged product of journalism, but a commodity controlled by market forces, and produced by all kinds of agencies and institutions.[2] Although information borrows the formats of news, it is not necessarily produced in a journalistic context. Information discourse echoes the style and features of journalistic narrative, such as quoting from experts, and thus presumably adopts the standards of objectivity and balance. Media formats and narrative structures are easily digestible by a large audience. Growing attention for the general audience has led to a commodification of images and imagination, as scientific public relations and merchandising drew attention to the presentation of scientific knowledge in popular media formats. Especially with large science projects, such as the Human Genome Project, an investment in products of popular culture is indispensable to help disseminate the tenets of genome mapping.

   Professional activities of journalists, scientists and public relations managers seem to converge because they are all purportedly involved in a communal project that overrides specific professional interests. An analysis of changing relations between scientists and journalists should not be read as a criticism of any group's professional practice. Rather, it divulges how professional speech communities invade each other's discursive terrain, while they continue to ascribe themselves professionally distinct responsibilities and authorities. The agency in charge of 'informing' the public of the ethical, legal and social consequences of the Human Genome Project is also in charge of promoting it. Journalists, like public relations managers, are expected to transmit scientific concepts to a general lay audience, while also critically investigating its claims. Control over popular images of genetics is part of a public relations strategy, from the earliest stage of design to its distribution in cultural products. Genome researchers are not at the mercy of journalists for crafting their public images, but play an essential role in the crafting process. Despite ardent insistence on professional division of labour, the 'image' of genetics is produced simultaneously by scientists, journalists and public relations managers. They continuously occupy each other's terrain, and all draw on the same discourse of information.[3] No analysis of the changing public image of genetics can proceed without awareness of the nature, origins and consequences of these shifting boundaries between the realms of science, journalism, public relations and fiction.

A final aspect of this level in the image-making process is the theatre's setting: the site or background against which the negotiation of images takes place. Setting is never a mere décor, a backdrop to embellish the actions of props, but constitutes a rhetorical resource in the configuration of genetics. Until the rise of the biotechnology industry, genetics was generally staged in the confines of the laboratory. The laboratory connoted either the sacred sanctuary of scientists, or represented to political activists a secluded bastion of power and arrogance. Images of the lab as the subterranean site of conspiracy, or contrastingly, as the secluded mecca for the practice of free scientific inquiry, both reflect the perception of the university lab as a public, scientific domain, where scientific knowledge can be freely pursued without any strings attached. The existence of the laboratory as a 'pure' space, untainted by social or commercial interests, is a utopian image, and it is as fictional as the idea of a subterranean space in which scientists conspire against humanity. With the advent of biotechnology, the locus of scientific action shifted from the university to the factory. Industrial parks located adjacent to university buildings betoken the fusion of science and industry. In popular representations, however, the 'untainted' image of genetic research taking place in public space remains pervasive. Genomics further diluted the marked boundaries between settings: genetic therapy required a clinical setting, and the activity of gene mapping symbolically moved genetics from a physical to a virtual setting.

Despite a dispersal of genetics research over industrial, university, clinical and virtual domains, the dominant setting becomes the hospital, providing an almost 'natural' backdrop for the combination of research and treatment. It is not an accidental site or locus, but reflects the emphasis on genetic therapy as the glorious triumph of curative medicine. Gene replacement therapy yields heroic images of young and helpless patients saved by the brave experiments of gene doctors. While medicine in the 1990s is a complex and contingent set of different, related practices – scientific, technological, clinical – the pre-eminent home for genomics presented in popular media is the hospital. A clinical setting, in contrast to a scientific or a business context, renders economics and politics irrelevant; it funnels public interest towards the saving of human life and away from potential consequences of new technologies or scientific concepts.

A more symbolic shift in the theatre setting over the years is the shift from physical to virtual space. Whereas in the 1960s and 1970s genetic research was frequently associated with the exploration of the moon and entire galaxies – reflecting Cold War political tensions – the HGP situates itself in the virtual space of the information highway, connoting a global scientific enterprise unhampered by national or political divisions. The new setting is the dynamic world of information flows, a virtual space which capitalizes on a constant exchange of digital data. Although this new setting invokes the idea of a unified, global scientific community working in the service of the human race, it tends to smooth over underlying animosity between national participants, which are almost exclusively American, Western European and Japanese. More importantly, the image effaces obvious economic interests in this global enterprise. Conceptualized as 'free space', genomics research is now regulated by economic laws, such as the ownership and control of genetic data instead of national or local laws. The 'free space' of virtual reality – erasing time and distance – and freely accessible to everyone, is as pastoral as the laboratory for unmitigated scientific inquiry. Both settings misleadingly convey the notion of science as uniquely serving the best interests of humankind.

Focusing on performance in the theatre of representation, I have tried to demonstrate how images and imagination are instrumental to the construction of popular knowledge. Images of genes, geneticists and genetics are constantly mobilized in the debate over the value of genetic engineering. Imagination, beyond merely reflecting general anxieties and fears about genetics, has also been a site of reconfiguration. From the production of images and imaginations we can discern how the dynamics between various professional groups are a constitutive element in the image-making process. The scripting, staging and setting of the theatre, rather than off-stage procedures, are part of the signification process. Whereas on the performance level we witnessed a peculiar incongruency between changing technologies and 'fixed' images and imaginations, on the level of production a similar parallax recurred. Various groups keep insisting on their distinct professional autonomy and discursive idiosyncrasy, while the boundaries between either special interest groups or professional groups appear notably fluid. The more fuzzy these boundaries get, the more adamant various groups seem to insist on their distinctiveness. In order to explain this recurrent paradox, it is illuminating to turn to the third level of analysis that the theatre metaphor allows: the transformation of culture as a backdrop for the production of genetic images.

THE TRANSFORMATION OF CULTURE

'The struggles that were once exclusively waged in the arena of production have, as a consequence, now spilled outwards to make of cultural production an arena of fierce social conflict', David Harvey concludes in his influential study on postmodernism.[4] There is no denying that genetics, as a public issue, has permeated the cultural and social fabric of society. The numerous stories in science, journalism and fiction have clearly shaped the popular representation of genetics. Writers have reacted ambivalently to this 'postmodern condition' of refracted scientific knowledge. Some theorists have lamented the 'narrativation' of knowledge, or the dissemination of knowledge through so many different discourses, and have located the source of this declivity in the changed conditions for communication. Jean-François Lyotard, among others, repeatedly bewailed the demise of 'grand narratives', or ideological theories that typically ordered our social strata and cultural apparatus.[5] Other theorists have welcomed the trend and pointed at the multiple discursive sites where resistance to dominant interpretations can materialize. Donna Haraway, for instance, has pointed at the many opportunities that the 'narrativation of knowledge' offers to affect the figuration of scientific knowledge.[6] The purported loss of the 'grand narratives', she argues, is rather a strategic regrouping of ideologies in the form of multiple stories.

The new local or 'molecular politics', as Fredric Jameson ironically characterizes the replacement of old-fashioned class and party politics, result in an absolute autonomy of the image or sign.[7] The postmodern era provides a cultural climate in which the image is no longer attached to a referent, but becomes a self-effacing reality. A condition for this change is the conflation of forms of high and mass culture, emanating from the logic of late capitalism. This transformation, in Jameson's view, has changed our very definition of culture:

> Culture itself falls into the world, and the result is not its disappearance but its prodigious expansion, to the point where culture becomes coterminous with social life in general: now all the levels become "acculturated," and in the society of the spectacle, the image, or the simulacrum, everything has at length become cultural (210).

Jameson's typification of postmodernist culture can be pinpointed in the materialization of the biotechnology industry and the Human Genome Project, where idea, image, product and production process collapse into a single signifier.

Although postmodernist theory clearly helps to signal certain cultural phenomena, it fails to provide a satisfactory gloss. In tracing popular images of genetics, I have not set out to prove either the validity or invalidity of the 'postmodern condition'. Rather than concluding that the dissemination of genetic knowledge has become thoroughly 'narrativized' and 'image contingent', I have wanted to use insights into the

transformation of culture to locate apparent frictions between the introduction of scientific or technological innovations and their popular representations. A cultural analysis of the 'imagenation' is as much concerned
with the cultural conditions that inscribe new technologies as with those
that reinscribe (old) ideologies. And it is at the junction of inscription and
re-inscription that we can notice a recurring paradox: the paradox between
increasingly hybrid technologies and scientific practices, and their monolithic or purified representations.

While the new genetics, and especially genomics, is motivating an
implosion of categories at various levels, the ontological categories that
distinguish the technical from the organic, the natural, and the textual are
vigorously reinstated.[8] The organic body coalesces with the digital body,
genes become interchangeable with bits and bytes, both reappearing as
'information'. The human genome – an abstract, engineered composite –
appears as the 'natural' body, an ideal to be attained through genetic
therapy. 'Edifying' metaphors appear in conjunction with vested interests
in 'authorship' and thus ownership. And while genome mapping is a
multinational enterprise, smoothly uniting corporate goals with public
interest, popular representations almost efface economic stakes. A search
for a universal 'code of codes' precludes notions of ambiguity; the practice of digital encoding prompts images of unilateral deciphering, hence
excluding multi-interpretability; and the 'natural' merger of science, economics and ethics in one conglomerate activity sails under the flag of
'public interest'. In other words, even though new concepts of genomics
orient themselves towards the constitution of a new order – a cyberculture
or technoculture – they are cemented in the well-known social and cultural
matrices of modernity.

The paradox of postmodern hybridization, however, is that it can apparently function only through a constant insistence on the distinction
between science and nature, science and representation, and science and
society. New concepts of flexibility and fluidity cannot exist without prior
notions of fixity. The idea that the physical (and mental) well-being of a
person is inscribed in the genes forms a neccesary preamble to the belief
that aberrant 'natural' patterns can be fixed. A standardized notion of a
genome as an abstract ideal inscribed in nature undergirds the notion that
technology can retool flawed bodies to attain this ideal. The image of a
coding standard resembles the ubiquitous media images of idealized
women's bodies. Model images inspire the development of cosmetic
surgery, even though these very images are the result of both physical
reconstruction and photographic retouching. By the same token, if genetic
predisposition – visualized by the 'DNA snapshot' – could not be held up

against a model of perfect health – the genome – the whole notion of genetic therapy would be up in the air. Both sides of the coin, fixity and fixability, derive from the idea(l) that we have identified as a science fiction staple, but which equally underlies the deployment and development of science: the optimization and perfection of the human body. A continuous emphasis on the optimization of the (biological) body funnels public attention to the 'wonders' of genetic surgery as a curative medical tool. 'Gene fixers', as the practitioners of gene therapy are also called, work on the replacement of human genes as if they were transplanting organs to save human lives. In the public mind, genetics coincides almost completely with genetic surgery, and the whole concept of genome mapping is conceived as subordinate to the ultimate goal of medically fixing imperfect bodies. Technology is needed to 'repair' natural flaws, and to conquer the 'secrets' of nature. Images of soldiers discovering and mapping lands figurally precede images of troops cultivating and mending natural resources. The fixed-yet-fixable hybrid departs from the notion that there is a distinct difference between nature and technology, as this distinction is vital to the whole enterprise of genetics.

In line with this ideology, the digital encoding of the human genome must be seen as a representation of the natural human body, since a code system carries meaning only in its (linear) relationship to the 'natural' organization of molecular units. Even though digital information also exists as an autonomous signifier, a self-referential system that can interrelate all kinds of information, most popular images impart a succinct one-to-one relationship between reality and representation. One can only 'own' and 'improve' strings of DNA-information if they are consolidated through a uniform relationship with real substance, just as one can only 'own' land if the symbolic inscription on a map has a referent in reality. Once again, the fluidity or 'autonomy' of representational systems functions only if assimilated with 'fixed' categories of reality and representations.

The incorporation of ethical, moral or feminist opposition into the geopolitics of human genome mapping is similarly overlaid with a strict enforcement of the boundaries between science and society. Although ethical and political activities are institutionalized in the 'scientific' structure of the HGP, they are not supposed to question the underlying tenets of genome research. Ethics are assumed to come into play as *consequences* of the implementation of technology, so that the boundaries between science and society are strictly upheld. And the fusion between the discourses of science, journalism, public relations and various forms of popular culture can only work through a constant insistence on the boundaries between the realms of fact, fiction and promotion. Hybridization can only succeed if paired off with insistence on the 'purity' of self-evident categories. Technological and scientific innovations facilitate the fluidity and flexibility of social domains, but are legitimated by dominant views of rigorously patrolled orders, only to make them appear as rational and efficient forms of practices.

Bruno Latour once wrote that postmodernism is a symptom, not an explanation.[9] Postmodernist theorists often explain manifestations of culture by pointing at a chronological and logical order between modernism and postmodernism, which presupposes that there has once been a modernist moment of epistemological and ontological purity. Symptomatic of postmodern culture – perhaps as symptomatic as hybridization and narrativation – is that it shares the contradictions of modernism, or, what Latour calls 'the upper and lower half of the modernist constitution': they bracket off nature from culture, science from society, and scientific practice from representation. While adherents to a modernist ideology accept the total division between the material and the technological world on the one hand, and the linguistic and textual world on the other, postmodern theorists often insist that every distinction between these categories has disappeared, and that the only reality left is the reality of the sign. To state that genomics is the result of a postmodern blurring of categories would be tautological; a more important diagnosis that can be derived from the analysis of the 'symptom' is that this blurring of categories gathers political and popular support through a systematic insistence on the separation of these categories. Or, in Latour's words, 'hybridization' cannot exist without 'purification'. An inquiry into modernity as the background for a current understanding of postmodern genomics raises important questions about destratification and deterritorialization of science and culture.

In analyzing popular images of genetics, I did not merely intend to show that the transformation of genetics reflects the general trend from a culture of modernity to postmodernity. 'Proving' such transformation, indeed, tacitly implies that there has ever been a pristine moment when the divisions between science and representation, science and nature, and science and society were clear-cut and incontrovertible. Neither is my reconstruction a critique of postmodern culture, in which the purported collapse of the genetic and the digital body leads to the theoretical orthodoxy that the body is a linguistic and discursive construction, and that genetics equals its constructed 'image'. In fact, I have demonstrated that there has never been a distinct separation between science and its images, just as there have never been 'purely' scientific and commercial domains, or public (academic) and private zones. Rather than reinstating those boundaries, it is important to recognize that traffic across various boundaries involves material as well as textual practices, financial as well as rhetorical resources, and material as well as symbolic objects.[10] Identification of a cultural subtext has helped explain why the emergence of technological innovations is often paradoxically paired off with the prolongation of traditional ideologies and hegemonic world views, such as genetic determinism. This inherent tension and cross-traffic proliferates most visibly through popular images and imaginations. More than anything, I have wanted to elucidate the connections between the immateriality of the image and the material conditions of its production.[11]

Images are never mere illustrations of scientific practices, and neither are imaginations mere reflections of people's anxieties in relation to science and technology. Images and imaginations, as I have stated earlier, are rhetorical tools in the construction of a public meaning. They are also intermediaries, establishing a link between nature and society, science and culture, reality and representation. It is only through such intermediaries that we can point out the interrelatedness and interconnectedness of seemingly unconnected phenomena arising from the same culture. As Marilyn Strathern has observed, 'culture consists in the images which make imagination possible, in the media with which we mediate experience'.[12] The 'grand narrative', or ideological belief in genetic determinism, runs so deep not because it is backed up by powerful technology, but because it is deeply rooted in the cultural images and imaginations instrumental to its dissemination. The 'images which make the imagination possible' are far from self-evident. The plot of Mary Shelley's *Frankenstein* does not necessarily lead to an interpretation of a technology that churns out monsters, and neither does the genome-as-book metaphor impart inevitable notions of deciphering. Images and imaginations, even if mythical and worn out, always allow for reinterpretation. Their stasis lies not in the images themselves, but in the way they are re-imagined, recycled through 'the media which mediates our experience'. It should not surprise us, at a time when technology promotes and facilitates the fluidity of information systems, that syntheses between domains of knowledge and modes of mediation emerge. Perhaps we should be more alarmed by the increasing push for 'fixed' images and the projection of rigid interpretive frameworks into the public consciousness. Equally important is the constant recognition of critical assessments of technology, the creative re-imaginings of cultural resources, and the questioning of commonsense culture.

Only if the transformation of genetics is interpreted against a backdrop of a changing culture, does it become clear how performances in the theatre of representation are staged, and how various groups of professionals contribute to the enactment and circulation of the scripted plays. At the beginning of this analysis, I introduced the metaphor of the theatre as a prism through which to assess the transformation of the public image of genetics. The theatre metaphor allows for a more complex picture of the dissemination or popularization of knowledge than previous models of linear information flows, such as the unidirectional vector of scientific information from experts to a general audience. The three layers of analysis account for an array of different, sometimes contradictory, images of genetics, for the politics of their mediation and for the cultural matrix in which they arise. A narrative analysis applied to these three levels has not only unveiled the complexity of the popularization process, but also the intricate connection between the 'image' and 'imagination' – melting together in the neologism 'imagenation'. Rather than a linear diffusion of knowledge, 'imagenation' assumes a recursive, circular transformation of knowledge. The notion of circularity captures both the complex structure of the genome and the multi-layered dissemination of genetic knowledge. The theatre metaphor does not provide a new model to theorize the mechanism of popularization, but provides an instrument for discerning the multiple manifestations of popular science.

Objections to the introduction of yet another metaphor in a field that is already dense with figurative language are not inconceivable. Am I trying to remedy an overdose of metaphors by throwing in another one? Perhaps so, but by introducing genetics as a 'theatre of representation' I have wanted to demonstrate more than the constitutive function of metaphors. I am not arguing that we are using *too many* images or imaginations in our evaluations and representations of science, but that we are not using our 'imagenation' *enough* in the reconfiguration of science, technology and society. Sometimes it seems as if we can hardly keep up with the images thrown at us, and that our imaginations can hardly keep up with the technology that is constantly changing our outlook on the way we can live our lives. 'Imagine having the tools to keep up with your imagination', states the text of an advertisement for Biosearch equipment.[13] Common sense tells us that imagination is always ahead of technology, and that our technological tools keep lagging behind. However, in the context of genomics, the opposite might be more accurate: our imaginative tools can hardly keep up with our technological innovations. Fortunately, imaginative tools are easier to access and use than technological ones, and the process of image-making is far more democratic than the process of scientific production. Given that everyone already participates in consuming and reconstructing science's perennial images and stories, the act of deciding what world we want from science is, *de facto*, everyone's business. And that, I think, is an empowering thought.

Notes

INTRODUCTION

1. H.G. Wells, *The Island of Dr Moreau*. New York: New American Library, 1977. Originally published 1896.
2. Movie, *The Island of Dr Moreau*. USA, director John Frankenheimer, 1996.
3. 'A Hollywood Production: Political Money' in *The New York Times* (12 September 1996) A1.
4. Movie, *Jurassic Park*. USA, director Steven Spielberg, 1993.

1 POPULAR IMAGES OF GENETICS

1. For a critical introduction into evolutionary biology, see Stephen Jay Gould, *Ever Since Darwin. Reflections in Natural History*. New York: Norton, 1977.
2. Hillary Rose, in *Love, Power, and Knowledge. Towards a Feminist Transformation of the Sciences* (Bloomington: Indiana University Press, 1994) Chapter 8, describes the term 'genetic turn'. See Sheldon Krimsky, in *Biotechnics and Society. The Rise of Industrial Genetics* (New York: Praeger, 1991) Chapter 1 for an explanation of the term 'geneticization'.
3. For an introduction into the relation between eugenics and the new genetics, see Troy Duster, *Backdoor to Eugenics* (New York: Routledge, 1990) particularly Chapters 6 and 7.
4. On the definition of 'public issues' and 'public problems' see Joseph Gusfield, *Drinking-Driving and the Symbolic Order* (Chicago: University of Chicago Press, 1981) Chapter 1.
5. In his excellent introduction to the history of genetics, Daniel Kevles traces the term 'genetic engineering' back to 1965. See *In the Name of Eugenics. Genetics and the Uses of Human Heredity* (Berkeley: University of California Press, 1985) pp. 264–8.
6. Cf. John Durant, Anders Hansen and Martin Bauer, 'Public Understanding of the New Genetics' in Theresa Marteau and Martin Richards (eds), *The Troubled Helix. Social and Psychological Implications of the New Human Genetics* (Cambridge: Cambridge University Press, 1996) pp. 235–48.
7. The history of eugenics is what makes deviance seem more problematic for many people than disease, as eugenic theories were used to disqualify people on the basis of inherited traits or intelligence. See, for instance, Mark H. Haller, *Myth of the Menace of the Feebleminded in Eugenics. Hereditarian Attitudes in American Thought*, New Brunswick: Rutgers University Press, 1963.
8. Cf. 'The XYY Man. Do Criminals Really Have Abnormal Genes?' in *Science Digest* (January 1976) pp. 32–8; 'Searching for a Gay Gene' in *Time* (12 June 1995) pp. 42–3; and 'Infidelity. It May Be in our Genes' in *Time* (15 August 1994) pp. 44–51.

9. Cf. J. Durant, G. Evans and G. Thomas, 'Public Understanding of Science' in *Nature* 340 (1989) pp. 11–14.

10. For a description of the public definition of infertility and the rise of in vitro fertilization as a remedy for infertility, see José Van Dijck, *Manufacturing Babies and Public Consent. Debating the New Reproductive Technologies* (London: Macmillan, 1995) Chapter 3.

11. See, for instance, John Erni, 'Articulating the (Im)possible: Popular Media and the Cultural Politics of "Curing Aids"' in *Communication* 13 (1992) pp. 39–56; and Edward Albee, 'AIDS. The Victim and the Press' in Thelma McCormack (ed.), *News and Knowledge* (London, Conn.: Jai Press, 1986) pp. 135–57.

12. See Brian Wynne, 'Public Understanding of Science Research: New Horizons or Hall of Mirrors?' in *Public Understanding of Science* 1 (1992) pp. 39–43. This article particularly questions the equation of 'public understanding' and 'public appreciation' of science.

13. On the notion of 'popularization equals pollution' see Jeremy Green, 'Media Sensation and Science. The Case of the Criminal Chromosome' in Terry Shinn and Richard Whitley (eds), *Expository Science: Forms and Functions of Popularisation* (Dordrecht: Reidel, 1985) pp. 139–61.

14. Cf. Maurice Goldsmith, *The Science Critic. A Critical Analysis of the Popular Presentation of Science* (London: Routledge, 1986) Chapter 1.

15. For a critical description of the 'diffusion model' and a proposal for an alternative model of 'translation', see Bruno Latour, *Science in Action. How to Follow Scientists and Engineers through Society* (Cambridge: Harvard University Press, 1987) pp. 132–41.

16. Cf. Bruce Lewenstein, 'The Meaning of Public Understanding of Science in the US after World War II' in *Public Understanding of Science* 1 (1992) pp. 45–68. Note that Lewenstein uses the term 'translation' where Latour uses the term 'diffusion'. See also Stephen Hilgartner, 'The Dominant View of Popularization. Conceptual Problems, Political Uses' in *Social Studies of Science* 20 (1990) pp. 519–39.

17. For an elaborate description of 'socially relevant groups', see Wiebe Bijker, *Of Bicycles, Bakelites, and Bulbs. Toward a Theory of Sociotechnical Change* (Boston: MIT Press, 1995) Chapter 2.

18. An interesting discussion of how scientists demarcate their rhetorical and professional terrain can be found in Charles A. Taylor, *Defining Science. A Rhetoric of Demarcation* (Madison: University of Wisconsin Press, 1996) Chapter 4.

19. See Bruno Latour and Steve Woolgar, *Laboratory Life. The Social Construction of Scientific Facts* (London: Sage, 1979); and Wiebe Bijker, Thomas Hughes and Trevor Pinch, *The Social Construction of Technological Systems. New Directions in the Sociology and History of Technology* (Cambridge: MIT Press, 1987).

20. Cf. Latour and Woolgar, *Laboratory Life*, p. 82.

21. It is impossible to separate images from facts, since the construction of images is always tied up with the construction of facts. Images, however, comprise those arguments that are implied, often alluded to, almost impalpable emotions that are elicited through various modes of expression, and as such they need to be distinguished from arguments or facts.

22. For a detailed introduction into a rhetorical analysis of science, and the terms logos, pathos and ethos, see Alan G. Gross, *The Rhetoric of Science* (Cambridge: Harvard University Press, 1990) Chapter 1.

23. A notable exception is Donna Haraway, who has always accounted for the specific impact of images of science in media and other forms of popular culture.

24. My notion of images as 'projections' or whipped up expectations of scientific 'products' does not equal actual 'promises' made by scientists or research and development departments. For a thorough analysis of the dynamics between scientists, research and development departments and marketeers, see Harro Van Lente, *Promising Technology. The Dynamics and Expectations in Technological Developments* (Delft: Eburon, 1993).

25. Cf. Marcel C. Folette, *Making Science our Own. Public Images of Science 1910–1950* (Chicago: University of Chicago Press, 1990) p. 4.

26. On the mutual shaping of technology and utopian visions, see for instance Howard P. Segal, *Future Imperfect. The Mixed Blessings of Technology in America* (Boston: University of Massachusetts Press, 1994) Chapter 1; for a specific case-study in the medical field, see Nancy Knight, '"The New Light". X-rays and Medical Futurism' in Joseph J. Corn (ed.), *Imagining Tomorrow. History, Technology, and the American Future* (Cambridge: MIT Press, 1986) pp. 10–34.

27. Rosalind Williams, *Notes on the Underground. An Essay in Technology, Society and the Imagination*. Cambridge: MIT Press, 1990.

28. I adopt Donna Haraway's approach to science as a 'story-telling practice', or in her own words 'a rule-governed, constrained, historically changing craft of narrating the history of nature'. *Primate Visions. Gender, Race and Nature in the World of Modern Science* (New York: Routledge, 1989) p. 4.

29. Cf. Joseph H. Gusfield, *Drinking-Driving and the Symbolic Order*, Chapter 1; and N. Katherine Hayles, 'Constrained Constructivism. Locating Scientific Inquiry in the Theater of Representation' in George Levine (ed.) *Realism and Representation* (Madison: University of Wisconsin Press, 1993) pp. 27–43.

30. In addition to theatre-related metaphors, I am also using a considerable number of military terms: strategies, mobilize, launch, wager, terrain, etc. Military metaphors have not only prevailed in scientific language of biomedicine, but have also been adopted by critics of technoculture. For a detailed analysis of military metaphors in biomedicine, see Scott L. Montgomery, 'Codes and Combat in Biomedical Discourse' in *Science as Culture* 2.12 (1991) pp. 341–90; and Bruno Latour, *Science in Action*, pp. 171–3.

31. Greg Myers argues that the discipline as a whole becomes an 'actor' in the scientific and popular narratives on genetics. Cf. Greg Myers, 'Making a Discovery. Narratives of Split Genes' in Christopher Nash (ed.), *Narrative in Culture. The Uses of Storytelling in the Sciences, Philosophy, and Literature* (London: Routledge, 1990) pp. 102–26.

32. Cf. Marcel C. Folette, *Making Science our Own*, Chapter 1.

33. Rosalynn D. Haynes, in *From Faust to Strangelove. Representations of the Scientist in Western Literature* (Baltimore: Johns Hopkins University Press, 1994), distinguishes six recurrent images of scientists in Western literature: the 'alchemist', the 'stupid virtuoso', the 'unfeeling' or arrogant scientist, the 'helpless scientist', the 'heroic adventurer' and the 'idealist'. These standard

images, according to Haynes, achieve significance as ideological indicators of a changing perception of science.

34. For an interesting analysis of photographs of scientists, see Daniel Jacobi and Bernard Schiele, 'Scientific Imagery and Popularized Imagery. Differences and Similarities in the Photographic Portraits of Scientists' in *Social Studies of Science* 19 (1989) pp. 731–53.

35. On the construction of facts and the use of plots in scientific prose, see Rom Harré, 'Some Narrative Conventions of Scientific Discourse' in Christopher Nash (ed.), *Narrative in Culture*, pp. 81–101; and Greg Myers, *Writing Biology. Texts in the Social Construction of Scientific Knowledge* (Madison: University of Wisconsin Press, 1990) Chapter 5.

36. *The Gene Race,* a co-production of BBC Horizon and PBS Nova, first broadcast 8 February 1996.

37. Mary Shelley, *Frankenstein or the Modern Prometheus*. London: Penguin Books, 1985. Originally published 1818.

38. For critical investigations into Frankenstein and its modern reinterpretations, see Fred Botting, *Making Monstrous. Frankenstein, Criticism, Theory* (Manchester: Manchester University Press, 1991); and George Levine and U.C. Knoepflmacher (eds), *The Endurance of Frankenstein* (Berkeley: University of California Press, 1979).

39. Aldous Huxley, *Brave New World*. New York: Harper and Row, 1946. Originally published 1932.

40. Huxley wrote *Brave New World* in reaction to Haldane's and Muller's revolutionary ideas on genetics, and the book was primarily meant to question the integrity and morality of scientists. Twenty-five years and a world war later, however, Huxley's mistrust of scientists had been largely replaced by a suspicion of politicians who implement scientific inventions for specific ideological purposes. In *Brave New World Revisited* (New York: Harper and Row, 1958), Huxley warns against the evil instruments of science that may easily fall into the hands of 'scientific dictators'. According to the author, the only mistake he made in his now classic tale of manipulated embryos grown in a lab to sustain a class society was that he had been too mellow in his predictions. The Nazi regime and the atom bomb are invoked to prove the author's conclusions that most of his speculations were overshadowed by real horrors.

41. Cf. Max Black, *Models and Metaphors* (Ithaca: Cornell University Press, 1966); and Mary B. Hesse and Max Black, *Models and Analogies in Science* (Notre Dame, Indiana: Notre Dame University Press, 1970).

42. See Gayle L. Ormiston and Raphael Sassower, *Narrative Experiments. The Discursive Authority of Science and Technology* (Minneapolis, University of Minnesota Press, 1989) p. 80; Sergio Sismondo, *Science without Myth. On Constructions, Reality, and Social Knowledge* (New York: SUNY Press, 1996) Chapter 8; and James J. Bono, 'Science, Discourse and Literature. The Role/Rule of Metaphor in Science' in Stuart Peterfreund (ed.), *Literature and Science. Theory and Practice* (Boston: Northeastern University Press, 1990) pp. 59–89.

43. George Lakoff and Mark Johnson, in *Metaphors We Live By* (Chicago: University of Chicago Press, 1980), have argued that metaphors can also create new meanings, resulting in a large and coherent network of entailments that edify our future experiences and ways of seeing.

44. On the notion of 'demetaphorization', see Sabine Maassen and Peter Weingart, 'Metaphors – Messengers of Meaning. A Contribution to an Evolutionary Sociology of Science' in *Science Communication* 17 (1995) pp. 9–31.

45. For an extensive description of the transformation of metaphors, see Michael A. Arbib and Mary B. Hesse, *The Construction of Reality* (Cambridge: Cambridge University Press, 1986) Chapter 8; and Lakoff and Johnson, *Metaphors We Live By*, Chapters 25–7.

46. Cf. Lewenstein, 'The Meaning of Public Understanding of Science', pp. 45–8; and June Goodfield, *Reflections On Science and Media* (Washington: AAAS Publication, 1981).

47. On the shifting perceptions of professional duties and responsibilities of journalists, see J. Herbert Altschull, *From Milton to McLuhan. The Ideas behind American Journalism* (New York: Longman, 1990) Parts VIII and IX.

48. Cf. Lakoff and Johnson, *Metaphors We Live By*, Chapter 20.

49. For a critical introduction to the ideas of postmodernism and the emancipation of the signifier, see David Harvey, *The Condition of Postmodernity* (Oxford: Basil Blackwell, 1989).

50. Jean-François Lyotard, *The Postmodern Condition. A Report on Knowledge*. Manchester: Manchester University Press, 1984.

51. For the term 'archaeology', I refer to Michel Foucault's suggestion to accept and 'follow the groupings that history suggests only to subject them at once to interrogation; to break them up and see whether they can be legitimately reformed' *The Archaeology of Knowledge* (New York: Pantheon, 1972) p. 26.

52. Michel Biezunski, 'Popularisation and Scientific Controversy. The Case of the Theory of Relativity in France' in Terry Shinn and Richard Whitley (eds), *Expository Science. Forms and Functions of Popularisation* (Dordrecht: Reidel, 1985) pp. 183–94.

53. Bruno Latour, *Science in Action*, Chapters 2 and 4.

2 BIOFEARS AND BIOFANTASIES

1. 'The Secret of Life' in *Time* (14 July 1958) pp. 50–54.

2. Cf. Evelyn Fox Keller, 'Making Gender Visible in the Pursuit of Nature's Secret' in Teresa de Lauretis (ed.), *Feminist Studies/Critical Studies* (Bloomington: Indiana University Press, 1986) pp. 67–77. In this essay, Keller contrasts the 'science of life' with the Manhattan Project symbolizing the 'secrets of death'.

3. In the 1930s, H.J. Muller already borrowed the authority of physics to establish molecular biology as a trustworthy science. Analogous to the materiality of the atom, Muller claimed DNA as the smallest physical entity of human life forms – the nucleus carrying all necessary information for the organism's future development. See Evelyn Fox Keller, *Secrets of Life, Secrets of Death. Essays on Language, Gender, and Science* (New York: Routledge, 1992) pp. 93–112.

4. 'Exploring the Secrets of Life' in *Newsweek* (13 May 1963) pp. 63–6.

5. Cf. Hans Blumenberg, *Die Lesbarkeit der Welt* (Frankfurt am Main: Suhrkamp, 1989).

6. Edwin Schrödinger, *What is Life? The Physical Aspects of the Living Cell*. Cambridge: Cambridge University Press, 1944.

7. For a detailed analysis of how metaphors of the phenotype were replaced by metaphors of the genotype, see Richard Doyle, 'Vital Language' in Carl F. Cranor (ed.), *Are Genes Us? The Social Consequences of the New Genetics* (New Brunswick: Rutgers University Press, 1994) pp. 52–68. I will return to Doyle's interesting discussion of metaphors in Chapter 6.

8. Edwin Chargaff, 'Vorwort zu einer Grammatik der Biologie. Hundert Jahre Nukleinsäureforschung' in *Experientia* 26 (1970) pp. 810–16.

9. Cf. Armand Mattelart and Michele Mattelart *Rethinking Media Theory. Signposts and New Directions* (Minneapolis: University of Minnesota Press, 1992) p. 28.

10. For a detailed analysis of the emergence of the gene as a master molecule, see Evelyn Fox Keller, 'Master Molecules' in Carl F. Cranor (ed.), *Are Genes Us?*, pp. 89–98.

11. 'Exploring the Secrets of Life' in *Newsweek* (13 May 1963) p. 64.

12. For an elaborate analysis of how the factory has been applied to the cell as well as higher organisms, see Sheldon Krimsky *Biotechnics and Society. The Rise of Industrial Genetics* (New York: Praeger, 1991) pp. 6–7. The factory as metaphor for the female reproductive system has been extensively discussed by Emily Martin, *The Woman in the Body. A Cultural Analysis of Reproduction* (Boston: Beacon Press, 1987) Chapters 3 and 4.

13. James Watson, *The Double Helix*. New York: Mentor, 1968.

14. Cf. Edward Yoxen, 'Speaking about Competition. An Essay on the Double Helix as Popularisation' in Terry Shinn and Richard Whitley (eds), *Expository Science. Forms and Functions of Popularisation* (Dordrecht: Reidel, 1985) p. 165.

15. For instance, Watson amusedly explains how he tricked Linus Pauling's son, their colleague, into giving him the information he needed to make sure that Pauling was on the wrong track. He also narrates triumphantly how he cajoled Maurice Wilkins into secretly giving him access to Rosalind Franklin's X-rays, which provided proof for the helical model.

16. Anne Sayre, *Rosalind Franklin and DNA. The First Full Account of Rosalind Franklin's Role in the Discovery of the Double-Helix Structure of DNA*. New York: Norton, 1975.

17. Cf. Hillary Rose, *Love, Power and Knowledge*, p.152.

18. Robert Olby, *The Path to the Double Helix* (London: Macmillan, 1974). Introduction by Francis Crick, p. v.

19. Movie, *Fantastic Voyage*. USA, director Richard Fleischer, 1996.

20. James D. Watson and Francis H.C. Crick, 'A Structure for Deoxyribose Nucleic Acid' in *Nature* (25 April 1953) p. 737.

21. As Alan Gross has elucidated, the two 'versions' of the discovery are both attempts at persuasion rather than information. The scientific article, in 1953, tried to argue to peers the existence of the double helix, while the autobiographical account in 1968 attempts to convince a larger audience of the 'truth' of this claim and the enormity of Watson's achievement. Cf. Alan Gross, *The Rhetoric of Science* (Cambridge: Harvard University Press, 1990) Chapter 4.

22. 'Man into Superman' in *Time* (19 April 1971) pp. 33–52.
23. *The Race for the Double Helix*, BBC Horizon series, broadcast 1974.
24. *Life Story*, BBC Horizon special, broadcast 1987.
25. For an interesting analysis of religious imagery and other symbols in this BBC production, see Sarah Franklin, 'Life Story. The Gene as Fetish and Object on TV' in *Science as Culture* 12 (1988) pp. 92–100.
26. On the use of religious inferences and symbolism in genetics in general, see Dorothy Nelkin and M. Susan Lindee, *The DNA Mystique. The Gene as a Cultural Icon* (New York: Freeman, 1995) Chapter 3.
27. See, for instance, June Goodfield, *Playing God. Genetic Engineering and the Manipulation of Life* (London: Hutchinson, 1977); Leroy Augenstein, *Come, Let Us Play God* (New York: Harper and Row, 1969); and Horace F. Judson, *The Eighth Day of Creation. Makers of the Revolution in Biology* (New York: Simon and Schuster, 1978).
28. 'Man into Superman. The Promise and Peril of the New Genetics' in *Time* (19 April 1971) pp. 33–52.
29. See, for instance, Vincent Detier, 'Christian Perspectives: Breaking the Genetic Code' in *Catholic World* 195 (August 1961) pp. 295–301.
30. 'Playing God' in *Newsweek* (23 November 1970) p. 54.
31. Robert T. Francoeur, *Utopian Motherhood. New Trends in Human Reproduction*. New York: A.S. Barnes, 1970.
32. Paul Ramsey, *Fabricated Man. The Ethics of Genetic Control*. New Haven: Yale University Press, 1970.
33. David Locke, *Science as Writing* (New Haven: Yale University Press, 1992) pp. 50–51.
34. 'Three Men and a Messenger' in *Time* (22 October 1965) p. 101.
35. Edward Yoxen, 'Speaking about Competition', p. 167.
36. Cf. Dorothy Nelkin, *Selling Science. How the Press Covers Science and Technology* (New York: Freeman, 1987) Chapter 1.
37. 'Exploring the Secrets of Life' in *Newsweek* (13 May 1963) pp. 63–6.
38. D.S. Halacy Jr., *Cyborg. Evolution of the Superman* (New York: Harper, 1965) p. 11.
39. Hermann J. Muller, in his famous book *Out of the Night. A Biologist's View of the Future* (New York: Vanguard Press, 1935), advocated the stocking of geniuses' semen in order to optimize the human gene pool in due time. Just what counts as a specimen of a superior human being appeared to be politically variable. In 1935, he suggested Lenin as an ideal sperm donor, but in later editions of his book, Lenin's name is conspicuously absent and replaced by names like Leonardo da Vinci, Descartes, Pasteur and Einstein. Halacy quotes Muller extensively to prove that the only reason why eugenics did not work at that time was that scientists lacked the actual instruments to improve upon human nature.
40. Gordon Rattray Taylor, *The Biological Time Bomb* (New York: World Publishing Co., 1968) Chapters 2 and 6.
41. Roughly outlined, the process of mononuclear reproduction entails the removal of the nucleus of an egg cell, leaving the body of the cell unharmed. Into the enucleated egg cell is put the nucleus of a body cell – not a sex cell – of the organism to be reproduced. An egg cell with forty-six chromosomes in its nucleus is then implanted in the uterus of a surrogate mother, where it

starts to grow into an exact duplicate of the donor of the cell tissue. The resulting child has neither a father nor a mother, only a donor of the nucleus of whom it is an exact genetic copy, with identical genes and chromosomes.

42. Cf. Donna Haraway, *Primate Visions,* p. 368.

43. The fact that in vitro fertilization was viewed as only a minor technical hurdle may be due to the optimistic note on which journalists reported every inch of progress in this field. In the public mind, IVF was already a reality before it actually occurred for the first time in 1978.

44. David Rorvik, 'Cloning. Asexual Human Reproduction?' in *Science Digest* 66 (November 1969) pp. 6–13.

45. Frank Herbert, *The Eyes of Heisenberg.* New York: Berkley Books, 1966.

46. In more than one respect, *The Eyes of Heisenberg* is reminiscent of Bulwer-Lytton's famous science fiction novel *The Coming Race* (1871). Just like Herbert's imagined perfect race of Optimen, Bulwer-Lytton's 'vril-ya' gained serenity and harmony through engineered evolution, but in the process lost the essential features of humanness: empathy, parental love and trust.

47. Ira Levin, *The Boys from Brazil.* London: Pan Books, 1976.

48. The description of the secret cloning of ninety-four genetic copies of Hitler is reminiscent of the 'bokanovskification' process described in Aldous Huxley's *Brave New World*. For details, see Susan Merrill Squier, *Babies in Bottles. Twentieth-Century Visions of Reproductive Technology* (New Brunswick: Rutgers University Press, 1994) pp. 146–8.

49. Nancy Freedman, *Joshua, Son of None.* London, Conn.: Granada, 1973.

50. G.K. Kellogg, a wealthy industrialist, was an active supporter of the American eugenics movement in the 1920s.

51. For an interesting discussion on how JFK's death shaped the collective memory and fantasy, see Barbie Zelizer, *Covering the Body. The Kennedy Assassination, the Media, and the Shaping of Collective Memory* (Chicago: University of Chicago Press, 1992).

3 BIOHAZARDS AND BIOETHICS

1. Michael Crichton, *The Andromeda Strain.* New York: Knopf, 1969; director Robert Wise, 1971.

2. Quite a few movies and books elaborate on the analogy between the invasion of aliens and the penetration of communist ideas, for instance *The Invasion of the Body Snatchers,* USA, director Don Siegel, 1956; and *It Came from Outer Space*, USA, director Jack Arnold, 1953.

3. S.E. Luria, 'Modern Biology. A Terrifying Power' in *The Nation* (20 October 1969) pp. 406–9.

4. Horace Freeland Judson, 'Fearful of Science' in *Harper's* (March 1975) pp. 32–41.

5. 'Genetic Moratorium' in *Time* (19 July 1976) p. 67; and 'Monstrous Microbes. The Dangers of Genetic Engineering' in *The Futurist* (October 1981) pp.17–21.

6. In 1972, the Environmental Protection Agency (EPA) came into existence; in 1970, the first Earth Day was organized by the environmental group Friends of the Earth.

7. Greg Bear, *Blood Music.* New York: Ace, 1985.

8. As some of the characters in *Blood Music* realize, past, present and future all merge in the noosphere. The infected Dr Bernard is able to re-create his past, make up with his estranged father, and redress a date with his first girlfriend. Time is only layers of the biological universe, which can be freely manipulated.

9. Cf. Sheldon Krimsky, *Genetic Alchemy. The Social History of the Recombinant DNA Controversy* (Cambridge: MIT Press, 1982) pp. 13–69. See also James D. Watson and John Tooze, *The DNA Story. A Documentary History of Gene Cloning* (San Francisco: Freeman, 1981) Chapters 3–5.

10. Robert Pollack, quoted from *Science* (9 November 1973) in Diana B. Dutton and Nancy E. Pfund, *Worse than Disease. Pitfalls of Medical Progress* (Cambridge: Cambridge University Press, 1988) p. 177.

11. Paul Berg *et al.*, 'Potential Biohazards of Recombinant DNA Molecules' in *Science* 185 (26 July 1974) p. 303.

12. Cf. Nicholas Wade, 'Genetic Manipulation. Temporary Embargo Proposed on Research' in *Science* 185 (26 July 1974) pp. 332–4. In his commentary, Wade assesses the effect and impact of Berg's letter published in the same issue of *Science*.

13. The three-day conference took place on 24–7 February, 1975 in Asilomar, California.

14. Paul Berg *et al.*, 'Asilomar Conference on Recombinant DNA Molecules' in *Science* (6 June 1975) pp. 991–4; National Institutes of Health, 'Recombinant DNA Research Guidelines' published in the *Federal Register* on 7 July 1976. See also Colin Norman, 'Genetic Manipulation: Guidelines Issued' in *Nature* 1 (July 1976) p. 2.

15. See, for instance, Dutton and Pfund, *Worse than Disease*, Chapter 6. See also Richard Hutton, *Bio-Revolution. DNA and the Ethics of Man-Made Life* (New York: New American Library, 1978) pp. 62–88; and Norton D. Zinder, 'A Personal View of the Media's Role in the Recombinant DNA War' in Raymond A. Zinlinskas and Burke K. Zimmerman (eds), *The Gene-Splicing Wars* (New York: Macmillan, 1986).

16. Cf. Dutton and Pfund, *Worse than Disease*, p. 180.

17. For a description of the impact of lawyers' lectures at Asilomar, see Hutton, *Bio-Revolution*, pp. 70–73.

18. For a detailed description of the press policy implemented by the Asilomar organizers, see Hutton, *Bio-Revolution*, pp. 80–86, and Zinder, 'A Personal View', p. 110.

19. The press invited to the Asilomar conference included reporters from *The New York Times, The Washington Post, The Los Angeles Times, The San Francisco Chronicle* and *Rolling Stone*.

20. For a general introduction to the changing mores in American journalism in the 1960s and 1970s, see Morris Dickstein, *Gates of Eden. American Culture in the Sixties* (New York: Penguin, 1977) pp. 150–53.

21. For analysis of the changing relations between various social movements and the press, see Todd Gitlin, *The Whole World is Watching. Mass Media and the Making and Unmaking of the New Left* (Berkeley: University of California Press, 1980); and Daniel C. Hallin, *The Uncensored War. The Media and Vietnam* (Berkeley: University of California Press, 1986). For a description of the press and the women's movement, see Gaye Tuchman, *Making News. A Study in the Construction of Reality* (New York: Free Press, 1978).

22. The 'new journalism' was propelled by Tom Wolfe in his introduction to the anthology *The New Journalism* (New York: Picador, 1974). New Journalism

208 *Notes*

has commonly been viewed as a stylistic innovation; see for instance Ronald Weber, *The Literature of Fact. Literary Nonfiction in American Writing* (Athens: Ohio State University Press, 1980); and John Hellmann, *Fables of Fact. The New Journalism as New Fiction* (Urbana Champaign: University of Illinois Press, 1980). For an interpretation of New Journalism as a cultural and political critique, see John Pauly, 'The Politics of New Journalism' in Norman Sims (ed.), *Literary Journalism in the Twentieth Century* (Oxford: Oxford University Press, 1990) pp. 110–30.

23. Michael Rogers, 'The Pandora's Box Congress' in *Rolling Stone* (19 June 1975) pp. 37–82. Rogers later published a more extensive account of the conference and its aftermath in *Biohazard* (New York: Knopf, 1977).

24. See, for instance, Martin S. Brander, 'The Scientist and the News Media' in *New England Journal of Medicine* (12 May 1983) pp. 1170–73.

25. For a detailed account of how various grassroots movements opposed the DNA-experiments, see Dutton and Pfund, *Worse than Disease*, pp. 189–91.

26. For a more elaborate review of the Cambridge controversy, see Watson and Tooze, *The DNA Story*, pp. 91–136.

27. The Cambridge guidelines were published in the *Bulletin of the Atomic Scientists* 33 (May 1977) p. 22.

28. Cf. Norman Zinder, 'A Personal View', p. 109.

30. The revised guidelines were published in the *Federal Register* on 22 December 1978.

31. Alan Gross regarded the intense public split among scientists over the regulation of DNA-experiments as a decisive stage in what he calls 'the social drama of recombinant DNA', *The Rhetoric of Science*, p. 186.

32. James D. Watson, 'In Defense of DNA' in *The New Republic* (25 June 1977) pp. 11–14.

33. As Watson insisted in *The DNA Story*: 'The matter was not only too technical but in a way also too fuzzy for responsibility to be easily shared with outsiders' (p. ix).

34. Rifkin and sympathizers from the People's Business Commission disrupted a forum organized by the NAS in Washington DC on 7–9 March 1977.

35. For instance, the EPA was successfully petitioned to regulate industry experiments with bioengineered organisms, and Rifkin also filed a suit against the Administration's regulatory policy on biotechnology, challenging its scientific basis, and requested an injunction against two scientists working on experiments involving oncogenes. Cf. Marjorie Sun, 'Rifkin Broadens Challenge in Biotech' in *Science* (20 July 1984) p. 297; Marjorie Sun, 'NIH Bows to Part of Rifkin Suit' in *Science* (30 November 1984) p. 1058; Marjorie Sun, 'Another Round in Rifkin versus Gene-Splicing' in *Science* (24 December 1984) pp. 1404–5; Marjorie Sun, 'Rifkin and NIH Win in Court Battle' in *Science* (15 March 1985) p. 1321; and William Booth, 'Of Mice, Oncogenes and Rifkin' in *Science* (22 January 1988) pp. 341–3.

36. Cf. 'The Most Hated Man in Science' in *Time* (4 December 1984) pp. 102–4.

37. Leslie Roberts, 'Ethical Questions Haunt New Genetic Technologies' in *Science* (3 March 1989) p. 1134.

38. See 'The Most Hated Man in Science' in *Time* (4 December 1984) pp. 102–4; and Horace Freeland Judson, 'Who Shall Play God?' in *Science Digest* (May 1985) pp. 52–91.

39. See for instance Jeremy Rifkin, 'One Small Step Beyond Mankind' in *The Progressive* (March 1977) p. 21.

40. An example of an 'advocacy journalist' account of the DNA debate is John Lear's *Recombinant DNA. The Untold Story* (New York: Crown, 1978). Lear's hypothesis that experiments with DNA could inadvertently lead to the creation of new pathogens is developed into a story whose plot-line is fairly simple: if 'we the people' unite against science, we will triumph over evil scientists.

41. Jeremy Rifkin, *Algeny*. New York: Viking Press, 1983.

42. In this respect, it is no coincidence that Rifkin was joined several times by religious groups, when petitioning Congress to ban genetic engineering of human reproductive cells.

43. 'Tinkering with Life' in *Time* (18 April 1977) pp. 32–45.

44. Arno G. Molutsky, 'Brave New World? Current Approaches to Prevention, Treatment, and Research of Genetic Diseases Raises Ethical Issues' in *Science* (23 August 1974) pp. 653–62.

45. Joseph Fletcher, *The Ethics of Genetic Control. Ending Reproductive Roulette* (New York: Anchor Press, 1974) pp. 127–8.

46. Cf. June Goodfield, *Playing God. Genetic Engineering and the Manipulation of Life* (London: Hutchinson, 1977) pp. 65–73.

47. Goodfield, *Playing God*, p. 199.

48. Amitai Etzioni, *The Genetic Fix* (New York: Harper, 1973) p. 37.

49. The recommendations of Etzioni and others led in 1980 to the instalment of a President's Commission for the Study of Ethical Problems in Medicine and Biomedical and Behavioral Research.

50. See, for instance, M. Hamilton (ed.), *The New Genetics and the Future of Man* (Grand Rapids: Eerdmans, 1972); and B. Hilton *et al.* (eds), *Ethical Issues in Human Genetics* (New York: Plenum, 1973).

51. David Rorvik, *In his Image. The Cloning of a Man*. New York: Lippincott, 1978.

52. Peter Gwynne, 'All About Clones' in *Newsweek* (20 March 1978) pp. 54–5.

53. June Goodfield, for instance, lamented that the increasing obfuscation between genres of fact and fiction jeopardized the credibility of journalists; journalists, according to her, have the obligation and responsibility to uphold this distinction. June Goodfield, *Science and the Media*. Washington: AAAS Publication, 1981.

54. Pamela Sargent, *Cloned Lives*. New York: Fawcett Gold Metal, 1976.

55. Caryl Rivers, 'Genetic Engineers: Now that they've gone too far, can they stop?' in *Ms* (June 1976) pp. 49–50.

56. Shulamith Firestone, *The Dialectic of Sex. The Case for a Feminist Revolution*. New York: Bantam, 1970.

57. One of the first science fiction novels tackling this issue was Charlotte B. Haldane's *Man's World* (1926), which depicts a society shaped by pre-natal sex selection. A reason for the popularity of this genre in addressing reproductive technologies, as Jenny Woolmark suggests, may be that science fiction offers a 'paraspace', a space in which 'subjectivity and experience, gender and identity, can be re-imagined in opposition to, and in recognition of, the dominant gendered discourses'. *Aliens and Others. Science Fiction, Feminism and Postmodernism* (Iowa City: Iowa University Press, 1994) p. 23.

58. Kate Wilhelm, *Where Late the Sweet Birds Sang*. New York: Harper and Row, 1976.
59. Marge Piercy, *Woman on the Edge of Time*. New York: Fawcett Crest, 1976.
60. Naomi Mitchison, *Solution Three*, London: Dobson, 1975.

4 BIOBUCKS AND BIOMANIA

1. Cf. Scott Montgomery, 'Codes and Combat in Biomedical Discourse' in *Science as Culture* 2:3 (1991) pp. 341–90. See also Paul De Kruif, *Microbe Hunters* (New York: Harcourt, Brace and Company, 1926).
2. Edward O. Wilson, *Sociobiology. The New Synthesis*. Cambridge: Harvard University Press, 1975.
3. Richard Dawkins, *The Selfish Gene*. Oxford: Oxford University Press, 1989. Originally published 1976.
4. 'The Secrets of the Human Cell' in *Newsweek* (20 August 1979) pp. 40–46.
5. 'DNA's New Miracles. How Science is Retooling Genes' in *Newsweek* (24 March 1980) pp. 44–9.
6. 'Getting Rich without a Product' in *Fortune* (16 June 1980) p. 149. 'Investors Dream of Genes' in *Time* (20 October 1980) p. 72.
7. Cf. Robert Teitelman, *Gene Dreams. Wall Street, Academia, and the Rise of Biotechnology* (New York: Harper, 1989) p. 4.
8. For an insightful analysis of the various meanings of the word 'biotechnology' see Paul Rabinow, *Making PCR. A Story of Biotechnology* (Chicago: University of Chicago Press, 1996) pp. 19–20.
9. Patricia Spallone has argued that the merger between eugenics and the new biology has manifested itself particularly in 'neutralized medical terms'. See *Beyond Conception. The New Politics of Reproduction* (Granby: Bergin and Garvey, 1989).
10. 'The Gene Doctors. Unlocking the Mysteries of Cancer, Heart Disease and Genetic Defects' in *Newsweek* (5 March 1984) pp. 34–40.
11. Dorothy Nelkin and Laurence Tancredi, *Dangerous Diagnostics. The Social Power of Biological Information* (New York: Basic Books, 1989) p. 9.
12. Nelkin and Tancredi, *Dangerous Diagnostics*, p. 102.
13. Geneticist Ananda Mohan Chakrabarty of General Electric Laboratory in Schenectady (NY) had applied for a patent on the germ 'Pseudomonas Aereoginosa' in 1972. The ensuing legal battle for the right to patent engineered bacterial strings took eight years. In 1980, the Supreme Court Justices ruled 5–4 that engineered organisms can be patented, thus giving the go-ahead to the commercial development of genetically engineered products.
14. Cf. Teitelman, *Gene Dreams*, pp. 27–35.
15. 'The Gene Doctors. Unlocking the Mysteries of Cancer, Heart Disease and Genetic Defects' in *Newsweek* (5 March 1984) pp. 34–40.
16. 'Shaping Life in the Lab. The Boom in Genetic Engineering' in *Time* (9 March 1981). On the cover of this *Time* issue, Genentech's Herbert Boyer is celebrated as the new prodigy of the biotech world.
17. See R.C. Lewontin, *Biology as Ideology. The Doctrine of DNA* (Concord, Ontario: Anansi Press, 1991); R.C. Lewontin, Steven Rose and Leon Kamin,

Not in Our Genes. Biology, Ideology and Human Nature (New York: Pantheon, 1984); and Jeffrey Weeks, *Sexuality and its Discontents. Meanings, Myths and Modern Sexualities* (London: Routledge, 1985) Chapter 5.

18. Renate D. Klein, 'Genetic and Reproductive Engineering. The Global View' in Jocelynne A. Scutt (ed.), *The Baby Machine. Reproductive Technology and the Commercialisation of Motherhood* (London: Merlin Press, 1988) pp. 235–72.

19. Gena Corea, 'Women, 'Class & Genetic Engineering. The Effects of New Reproductive Technologies on All Women' in Jocelynne A. Scutt (ed.), *The Baby Machine*, pp. 135–56.

20. See, for instance, Janice G. Raymond, *Women as Wombs. Reproductive Technologies and the Battle over Women's Freedom* (New York: Harper Collins, 1993); and Robyn Rowland, *Living Laboratories. Women and Reproductive Technologies* (Bloomington: Indiana University Press, 1992).

21. Hillary Rose, *Love, Power, and Knowledge*, p. 186.

22. 'Where Genetic Engineering Will Change Industry' in *Business Week* (22 October 1979) pp. 160–72; and 'Investor's Dream of Genes. Genentech, a Pioneer in Daring DNA Research, Goes Public' in *Time* (20 October 1980) p. 72.

23. 'Scrambling for Gene' in *Newsweek* (27 October 1980) p. 54.

24. By 1985, investments in the biotech industry topped 2.5 billion dollars. See Dutton and Pfund, *Worse than Disaese*, p. 210. On the rapid growth of the biotech industry, see also Teitelman, *Gene Dreams*, Chapters 4–10.

25. Spyros Andreopoulos, 'Gene Cloning by Press Conference' in *New England Journal of Medicine* 304: 13 (27 March 1980) pp. 743–6.

26. Barbara J. Culliton, 'Biomedical Research Enters the Marketplace' in *New England Journal of Medicine* 304: 20 (14 May 1981) pp. 1195–201.

27. Quoted in John Elkington, *The Gene Factory. Inside the Genetic and Biotechnology Business Revolution* (New York: Caroll and Graff, 1985) p. 59.

28. Cetus Corporation, Annual Report 1981, p. 2.

29. 'Getting Rich without a Product' in *Fortune* (14 June 1980) p. 149; and 'If Nothing Else, Genetic Engineering Can Turn a Scientist into a Multimillionaire' in *People* (29 December 1980) pp. 32–3.

30. *New Scientist* (21 May 1987) pp. 38–71. The advertisements do not have page numbers, but are interspersed between the articles.

31. Cf. Dutton and Pfund, *Worse than Disease*, p. 205.

32. Cf. R.C. Lewontin, 'The Dream of the Human Genome' in the *New York Review of Books* (28 May 1992) pp. 31–9.

33. Cf. Teitelman, *Gene Dreams*, Chapters 6 and 7.

34. 'Biotechnology. Capitalizing on Life' in *Science Digest* (June 1986) pp. 32–79.

35. The Congressional hearings were held in 1981–2 and were led by Albert Gore. A number of scientific journals devoted editorials or columns to the issue of commercialization of genetics research. See, for instance, 'Should Academics Make Money Outside?' (editorial) in *Nature* 286 (24 July 1980) 319–20.

36. For a detailed description of the outcome of the Pajaro Dunes Conference, see Martin Kenney, *Biotechnology. The University-Industrial Complex* (New Haven: Yale University Press, 1986) pp. 85–9.

37. Cf. 'Statement Issued after the Pajaro Dunes Conference' reprinted in *TechTalk* (7 April 1982) p. 8.
38. Edward Yoxen, *The Gene Business. Who Should Control Biotechnology?* (New York: Harper and Row 1983) p. 48.
39. Martin Kenney, *Biotechnology*, pp. 1–58.
40. For instance, Harvard University struck up a long-term co-operation agreement with chemical company DuPont, and the University of California, at Davis, teamed up with Allied Corporation.
41. 'Genes For Profit' in *Science Digest* (May 1982) pp. 12–13.
42. 'DNA Research Gets Down to Business' in *The Nation* (13 October 1979) pp. 326–7.
43. 'Scrambling for Biotech Bucks' in *The Nation* (10 April 1989) pp. 476–8.
44. For a more detailed description of this incident, see Chapter 5 of this book.
45. Stephen Hill and Tim Turpin, 'Academic Research Cultures in Collision' in *Science as Culture* 3: 4 (1994) 327–61.
46. Edward Yoxen, *The Gene Business*, p. 190.
47. For a typical defence of the merger between industry and academia, see Arthur Kronberg, *The Golden Helix. Inside Biotech Ventures* (Sausalito: University Science Books, 1995).
48. For an interesting analysis of 'science parks' in Great Britain, see Doreen Massey, Paul Quintas and David Wield, *High-Tech Fantasies: Science Parks in Society, Science, and Space* (London: Routledge, 1992).
49. 'Beyond the Double Helix' in *Esquire* (November 1985) pp. 196–210.
50. Stephen Hall, *Invisible Frontiers. The Race to Synthesize a Human Gene*. Washington: Microsoft Press, 1987.
51. Robin Cook, *Mutation*. New York: Berkley Books, 1989.
52. For a survey of Frankenstein re-creations in film, see Albert J. Lavalley, 'The Stage and Film Children of Frankenstein: A Survey' in Levine and Knoepflmacher, *The Endurance of Frankenstein*, pp. 243–88.
53. For feminist interpretations of Mary Shelley's Frankenstein, see, for instance, Ellen Moers, 'Female Gothic' in Levine and Knoepflmacher (eds), *The Endurance of Frankenstein*, pp. 77–87; Mary A. Favret, 'A Woman Writes the Fiction of Science: The Body in Frankenstein' in *Genders* 14 (Fall 1992) pp. 50–65; Anne K. Mellor, *Mary Shelley. Her Life, her Fiction, her Monsters* (New York: Routledge, 1988); and Mary K. Patterson Thornburg, *The Monster in the Mirror* (Ann Arbor: University of Michigan Press, 1987).
54. Fredric Jameson, 'Postmodernism, or the Cultural Logic of Late Capitalism' in *New Left Review* 146 (1984) pp. 53–92.

5 BIOPHORIA: THE HUMAN GENOME PROJECT

1. R. Edgar and H. Noller, *Human Genome Institute: A Position Paper*. Biology Board of Studies, University of California, Santa Cruz, 31 October 1984. Quoted by Robert Cook-Deegan, *The Gene Wars. Science, Politics, and the Human Genome* (New York: Norton, 1994) p. 82.
2. The 'Five-Year-Plan', *Understanding our Genetic Inheritance. The U.S. Genome Project 1991–1995* was submitted to and approved by Congress in 1990.

3. Evelyn Fox Keller, *Refiguring Life. Metaphors of Twentieth-Century Biology* (New York: Columbia University Press, 1995) p. 89.

4. Cf. 'Proposal to Sequence the Human Genome Stirs Debate' in *Science* 232 (27 June 1986) pp. 1598–900.

5. *Human Genome News* (September 1990) pp. 2 and 3.

6. See Paul Rabinow, who explains how in the age of gene 'mapping' the pathological will be defining the normal. Paul Rabinow, 'Artificiality and Enlightenment: From Sociobiology to Biosociality' in Jonathan Crary and Sanford Kwinter, *Incorporations* (New York: Zone, 1992) pp. 234–52.

7. 'Seeking a Godlike Power', special issue of *Time* (Fall 1992).

8. 'Genetics. The Future is Now' in *Time* (17 January 1994) pp. 40–51.

9. Walter Gilbert, 'A Vision of the Grail' in Daniel J. Kevles and Leroy Hood, *The Code of Codes. Scientific and Social Issues in the Human Genome Project* (Cambridge: Harvard University Press, 1992) pp. 83–97.

10. For an excursion into the differences between analog and digital recording, see Mark Poster, *The Mode of Information. Poststructuralism and Social Context* (Chicago: University of Chicago Press, 1990) Chapter 3.

11. The composite of which the human genome is made is referred to as the 'equivalent to the unknown soldier'. *Human Genome News* (July 1990) p. 5.

12. Cf. Donna Haraway, 'The Biopolitics of Postmodern Bodies: Determinations of Self in Immune Discourse' in *Differences* 1 (Winter 1989) pp. 3–43; and Donna Haraway, *Simians, Cyborgs and Women. The Reinvention of Nature* (London: Free Association Books, 1991). See also Rabinow, 'Artificiality and Enlightenment', p. 241.

13. Keller, *Refiguring Life*, p. 117.

14. For a detailed description of the discussion on patents and copyrights of the human genome, see Joel Davis, *Mapping the Code. The Human Genome Project and the Choices of Modern Science* (New York: Wiley and Sons, 1990) Chapter 6.

15. *Human Genome News* (March 1991) p. 7.

16. See, for instance, William Cookson, *The Gene Hunters. Adventures in the Genome Jungle* (London: Aurum Press, 1994).

17. 'The Gene Hunt' in *Time* (20 March 1989) pp. 69–71.

18. Lois Wingerson, *Mapping our Genes: The Genome Project and the Future of Medicine* (New York: Penguin, 1990) p. 299.

19. 'The Gene Kings' in *Business Week* (8 May 1995) pp. 72–8.

20. Cf. Jerry E. Bishop and Michael Waldholz, *Genome. The Story of our Astonishing Attempt to Map All the Genes in the Human Body* (New York: Simon and Schuster, 1990) p. 219. Another example of the Lewis and Clark analogy can be found in Cook-Deegan, *The Gene Wars*: 'Botstein derided the notion of genome sequencing, noting that Lewis and Clark had followed a similar approach to mapping the American West, a millimeter at a time, they would still be somewhere in North Dakota' (111).

21. Quoted in 'The Gene Hunt' in *Time* (20 March 1989) p. 70.

22. Walter Gilbert, 'A Vision of the Grail', p. 96 (emphasis added).

23. 'Seeking a Godlike Power' in *Time*, special issue (Fall 1992) p. 58.

24. 'The Age of Genes' in *U.S. News and World Report* (4 November 1991) pp. 64–76.

25. 'Genetics, the Future is Now' in *Time* (17 January 1994) p. 42.

26. 'The Genome Project' in *The New York Times Magazine* (13 December 1987) p. 44.

27. Maya Pines, *Mapping the Human Genome* (Bethesda, Maryland: Howard Hughes Medical Institute, 1987) p. 3.

28. Cf. Evelyn Fox Keller, *Secrets of Life, Secrets of Death*, p. 41.

29. *Human Genome News* (January 1992) p. 3.

30. *Human Genome News* (September 1993) p. 3.

31. 'Riding the DNA Trail' (profile of Francis Collins) and 'Battler for Gene Therapy' (profile of French Anderson) in *Time* (17 January 1994) pp. 48–51.

32. 'Stopping Cancer in its Tracks' in *Time* (25 April 1994) pp. 40–48.

33. Cf. T. Hugh Crawford, 'Imaging the Human Body: Quasi Objects, Quasi Texts, and the Theater of Proof' in *PMLA* 3: 1 (January 1996) pp. 66–79.

34. Cook-Deegan, *The Gene Wars*, p. 122.

35. See, for instance, 'Genome Project Goes Commercial' in *Science* 259 (15 January 1993) pp. 300–2. And 'A Showdown over Gene Fragments' in *Science* 266 (14 October 1994) pp. 208–10.

36. 'Genes Jump into the Marketplace' in *The Lancet* (11 May 1996) p. 1323.

37. Louis Harris and Ass., March of Dimes *Survey of Attitudes of American Adults toward Using Gene Therapy*. This poll, conducted and published in 1992, showed that 87 per cent of American adults approve of gene transfer.

38. Larry Thompson, *Correcting the Code. Inventing the Genetic Cure for the Human Body*. New York: Simon and Schuster, 1994.

39. UCLA scientist Martin J. Cline, back in 1980, had proposed an experiment with gene transfer, but his application was rejected. Cline, eager to evade laborious protocols and reluctant committees, then went to Israel and Italy to conduct his experiments. He failed miserably, according to Thompson, not just because his methods did not work scientifically, but because he could never have won public approval for the experiment. The public, in 1980, had just barely recovered from seven years of messy battles about the dangers of recombinant DNA-research, and was not ready for this experiment (*Correcting the Code*, pp. 189–217).

40. Thompson, *Correcting the Code*, pp. 329–35.

41. *The Dream Continues*. Genentech Annual Report. San Fransico: Genentech Inc., 1993.

42. *The Dream Continues*, p. 15.

43. *Nature Genetics* 8: 2 (October 1994): advertisements do not have page numbers.

44. *Nature Genetics* 11: 1 (September 1995); and *Nature Genetics* 6: 1 (January 1994).

45. *Nature Genetics* 6: 2 (February 1994).

46. In 1992, the WGBH Educational Foundation received support from the NCHGR to produce *The Secret of Life*, a series aired on public broadcasting in September 1993. *Medicine at the Crossroads* is a four-part documentary on genome mapping produced by WNET in Boston, with the financial help of the NIH.

47. *The Human Genome Project*, public relations video produced by the Public Affairs Department of the NCHGR, Bethesda, Maryland, 1991.

48. Advertisement for New England Biolabs in *Nature Genetics* 8: 2 (October 1994).

49. See, for instance, *Human Genome News* (September 1993), where we can read that AmpliTaq Inc. generously allows scientists to purchase large quantities of an enzyme at a reduced rate.
50. G.J.V. Nossal and Ross L. Coppel, *Reshaping Life. Key Issues in Genetic Engineering* (Cambridge: Cambridge University Press, 1989) Chapters 10 and 11.
51. See, for instance, David Suzuki and Peter Knudtson, *Genethics. The Clash between the New Genetics and Human Values*. Cambridge: Harvard University Press, 1990.
52. For a description of ELSI's goals, see Department of Health and Department of Energy, *Understanding our Genetic Inheritance. The U.S. Human Genome Project: The First Five Years 1991–1995* (NIH Publication, April 1990) pp. 20–2.
53. See, for instance, Jon Turney, 'Thinking about the Human Genome Project' in *Science as Culture* 19 (1993) pp. 282–94.
54. This interpretation is underscored by the news that Lori Andrews, chair of a working group set up jointly by the NIH and the DOE to advise the ELSI committee, resigned. In her resignation, Andrews cited concern about the autonomy of the group, and expressed frustrations about frequent interference by the NCHGR. See 'Genome Ethics Chair Resigns Amid Worries over Autonomy' in *Nature* 380 (14 March 1996) p. 96.
55. 'Hunting Down Huntington's' in *Discover* (December 1993) pp. 99–107; and 'To Catch a Killer Gene' in *New Scientist* (24 April 1993) pp. 37–41.
56. For a more elaborate description of ELSI's institutionalization, see José Van Dijck, 'The Human Genome Narrative' in *Science as Culture* 23 (1995) pp. 217–47.
57. The 'ethics of care' philosophy was first propelled by Carol Gilligan in *In a Different Voice* (Cambridge: Harvard University Press, 1982). For specific applications of Gilligan's theory to medical ethics, see Virginia L. Warren, 'Feminist Directions in Medical Ethics' in *Hypathia* 4: 2 (Summer 1989) pp. 31–45; and Susan Sherwin, 'Feminist and Medical Ethics: Two Different Approaches to Contextual Ethics' in *Hypathia* 4: 2 (Summer 1989) pp. 17–29. For criticism of the 'ethics of care' approach, see Kathy Davis, 'Towards a Feminist Rhetoric: The Gilligan Debate Revisited' in *Women's Studies International Forum* 15: 2 (1992) pp. 219–31, and Katha Pollitt, 'Are Women Morally Superior than Men?' in *The Nation* (28 December 1992) pp. 799–807.
58. For an extensive critique of how genomics or bio-informatics has ungendered the female body, see, for instance, Barbara Stafford, *Body Criticism. Imaging the Unseen in Enlightenment Art and Medicine* (New York: Zone, 1991) pp. 1–40; and Allucquere Rosanne Stone, 'Will the Real Body Please Stand Up? Boundary Stories about Virtual Culture' in Michael Benedikt, *Cyberspace. First Steps* (Cambridge: MIT Press, 1991) pp. 81–118.
59. Karen Schmitt and P.N. Goodfellow, 'Predicting the Future' in *Nature Genetics* 7 (June 1994) p. 219.
60. *Human Genome News* (September 1993) p. 2.
61. Web site address: Http://www.ncbi.nlm.nih.gov/science96/.
62. 'The Race to Map Our Genes' in *Time* (8 February 1993) p. 57.

6 BIOCRITICISM AND BEYOND

1. Thomas Fogle, 'Information Metaphors and the Human Genome Project' in *Perspectives in Biology and Medicine* 38: 4 (Summer 1995) p. 535.
2. Richard Doyle, 'Vital Language' in Carl F. Cranor (ed), *Are Genes Us? The Social Consequences of the New Genetics* (New Brunswick: Rutgers University Press, 1994) pp. 52–68.
3. Bonnie B. Spanier, *Im/partial Science. Gender Ideology in Molecular Biology* (Bloomington: Indiana University Press, 1995) pp. 149–50.
4. Mary Rosner and T.R. Johnson, 'Telling Stories: Metaphors of the Human Genome Project' in *Hypathia* 10: 4 (Fall 1995) pp. 104–29.
5. Rosner's and Johnson's concretization of this alternative 'cyborg practice', however, remains vague, as they merely point at Donna Haraway as a trend-setter in cyborg reading, but do not provide any specific examples of alternative genomics images or readings.
6. Robert Pollack, *Signs of Life. The Language and Meanings of DNA* (New York: Houghton Miffkin, 1994) p. 5.
7. Richard Powers, *The Gold Bug Variations*. New York: Harper, 1991.
8. For an interesting analysis of musical theory in *The Gold Bug Variations*, see Jay A. Labinger, 'Encoding an Infinite Message: Richard Powers's *Gold Bug Variations*' in *Configurations* 3 (1995) pp. 79–94.
9. Dorothy Nelkin and M. Susan Lindee, *The DNA Mystique. The Gene as a Cultural Icon* (New York: Freeman, 1995) p. 2.
10. For a more detailed description of the changing definition of motherhood, see José Van Dijck, *Manufacturing Babies and Public Consent* (London: Macmillan, 1995) Chapter 6; and Valerie Hartouni, 'Breached Births: Reflections on Race, Gender and Reproductive Discourse in the 1980s' in *Configurations* 1 (1994), pp. 73–88.
11. Ruth Hubbard and Elijah Wald, *Exploding the Gene Myth. How Genetic Information is Produced and Manipulated by Scientists, Physicians, Employers, Insurance Companies, Educators, and Law Enforcers* (Boston: Beacon Press, 1993). For the remainder of this chapter, I will refer to Ruth Hubbard as the single author of this book because she claims responsibility for the scientific content in the preface.
12. This does not mean, by the way, that most detractors of genomics abstain from political action. On the contrary, scientists like Ruth Hubbard have been actively involved in grassroots movements and political action groups.
13. Cf. Bruno Latour, 'Drawing Things Together' in Steve Woolgar and Michael Lynch (eds), *Representation in Scientific Practice* (Boston: MIT Press, 1990) pp. 19–67.
14. M. Mitchell Waldrop, 'On-line Archives Let Biologists Interrogate the Genome' in *Science* 269 (8 September 1995) p. 1358. The use of the Visual Human Data has been licensed to over 300 sites, where it has been mostly applied as a multi-media textbook of anatomy.
15. Cf. Mitchell Waldrop, 'The Visible Man Steps Out' in *Science* 269 (8 September 1995) p. 1358.
16. Kevin Clarke's web site address: Artnetweb:http://artnetweb.com/artnetweb/projects/clarke/kchome1.html.

17. As one critic comments on Nell Tenhaaf's work: 'Hers is a strategy of occupation rather than repudiation, one that suggests we can exercise critical judgement and work towards social change by inhabiting these territories.' Kim Sawchuk, 'Biological. Not Determinist. Nell Tenhaaf's Technological Mutations' in *Parachute* 75 (July/August 1994) pp. 11–17.
18. Nell Tenhaaf, 'Mutational Cravings' in *C* (Winter 1993) pp. 46–51.
19. Amy Thomson, *Virtual Girl*. New York: Ace, 1993.
20. Octavia Butler, *Xenogenesis*. New York: Warner Books. Part 1: *Dawn*, 1987. Part 2: *Adulthood Rites*, 1988; Part 3: *Imago*, 1989.

7 RETOOLING THE IMAGINATION

1. My use of the term 'positioning' covers Donna Haraway's notion of 'situated knowledge'. All responses to genetics, whether from scientists or feminists, are articulated from a specific interested position. Haraway particularly opposes the idea that there is an 'outside text' or a transcendent meta-viewpoint from which scientific constructions can be theorized. Cf. 'Situated Knowledges. The Science Question in Feminism and the Privilege of the Partial Perspective' in *Feminist Studies* 14: 3 (Fall 1988) pp. 575–99; and *Simians, Cyborgs and Women*, p. 81.
2. Mark Poster has characterized this phenomenon as the transition from the 'mode of presentation' to the 'mode of information'. Poster models this distinction after Richard Terdiman's theory of the gradual change, in the nineteenth century, from a contextualized, linear presentation of newspaper stories to a collage of isolated data and stories, giving an appearance of objectivity because the subjective 'collector' or mediator of facts disappeared from the scene. Cf. *The Mode of Information*, p. 62.
3. For a theoretical exploration of demarcation dynamics between professional groups, see Charles A. Taylor, *Defining Science,* Chapter 1.
4. David Harvey, *The Condition of Postmodernity*, p. 61.
5. As Lyotard states: 'We may form a pessimistic impression of this splintering: nobody speaks all of those languages, they have no universal meta-language … the diminished task of research have become compartmentalized and no one can master them all… That is what the postmodern world is all about. Most people have lost nostalgia for the lost narrative.' *The Postmodern Condition*, pp. 38–41.
6. Cf. Donna Haraway, who argues that the 'diffraction' of science into multiple narratives also offers multiple sites for reconfiguration (*Primate Visions*, p. 370).
7. Fredric Jameson, 'Periodizing the Sixties' in Sohnya Sayres *et al.*, *The Sixties without Apology* (Minneapolis: University of Minnesota Press, 1984) p. 210.
8. Arthur Escobar argues that techno-science is motivating a blurring and implosion of categories that have defined the natural, organic, technical and textual: 'While nature, bodies and organisms certainly have an organic basis, they are increasingly produced in conjunction with machines, and this production is always mediated by scientific narratives (discourses of biology, technology and the like) and by culture in general.' Arthur Escobar,

'Welcome to Cyberia. Notes on the Anthropology of Cyberculture' in *Current Anthropology* 35: 3 (June 1994) pp. 211–31.

9. Cf. Bruno Latour, *We Have Never Been Modern* (Cambridge: Harvard University Press, 1993) p. 131.

10. For an illuminating introduction in the cross-boundary practices of the cultural studies of science, see Joseph Rouse, 'What Are the Cultural Studies of Scientific Knowledge?' in *Configurations* 1 (1992) pp. 1–22.

11. N. Katherine Hayles, 'The Materiality of Informatics' in *Configurations* 1 (1992) pp. 147–70.

12. Marilyn Strathern, *Reproducing the Future. Anthropology, Kinship, and the New Reproductive Technologies* (Manchester: Manchester University Press, 1993) p. 32.

13. Advertisement for Biosearch Labs, in *Science* 265 (30 September 1994): no page number.

Bibliography

Albee, Edward. 'AIDS. The Victim and the Press.' *News and Knowledge.* Ed. Thelma McCormack. London (Conn.): Jai Press, 1986. 135–57.

Altschull, Herbert. *From Milton to McLuhan. The Ideas behind American Journalism.* New York: Longman, 1990.

Arbib, Michael A. and Mary B. Hesse. *The Construction of Reality.* Cambridge: Cambridge University Press, 1986.

Augenstein, Leroy. *Come, Let Us Play God.* New York: Harper, 1969.

Bear, Greg. *Blood Music.* New York: Ace, 1985.

Benedikt, Michael. *Cyberspace. First Steps.* Cambridge: MIT Press, 1991.

Biezunski, Michel. 'Popularisation and Scientific Controversy. The Case of the Theory of Relativity in France.' *Expository Science: Forms and Functions of Popularisation.* Eds Terry Shinn and Richard Whitley. Dordrecht: Reidel, 1985. 183–94.

Bijker, Wiebe E. *Of Bicycles, Bakelites, and Bulbs. Toward a Theory of Sociotechnical Change.* Boston: MIT Press, 1995.

Bijker, Wiebe, Thomas Hughes and Trevor Pinch. *The Social Construction of Technological Systems. New Directions in the Sociology and History of Technology.* Cambridge: MIT Press, 1987.

Bishop, Jerry E. and Michael Waldholz. *Genome. The Story of our Astonishing Attempt to Map All the Genes in the Human Body.* New York: Simon and Schuster, 1990.

Black, Max. *Models and Metaphors.* Ithaca: Cornell University Press, 1966.

Blumenberg, Hans. *Die Lesbarkeit der Welt.* Frankfurt am Main: Suhrkamp, 1989.

Bono, James J. 'Science, Discourse and Literature. The Role/Rule of Metaphor in Science.' *Literature and Science.Theory and Practice.* Ed. Stuart Peterfreund. Boston: Northeastern University Press, 1990, 59–89.

Botting, Fred. *Making Monstrous. Frankenstein, Criticism, Theory.* Manchester: Manchester University Press, 1991.

Butter, Octavia. *Xenogenesis.* New York: Warner Books. Part One: *Dawn*, 1987; Part Two: *Adulthood Rites*, 1989; Part Three: *Imago*, 1989.

Chargaff, Edwin. 'Vorwort zu einer Grammatik der Biologie. Hundert Jahre Nukleinsäureforschung.' *Experientia* 26 (1970): 810–16.

Cook, Robin. *Mutation.* New York: Berkley, 1989.

Cook-Deegan, Robert. *The Gene Wars. Science, Politics, and the Human Genome.* New York: Norton, 1994.

Cookson, William. *The Gene Hunters. Adventures in the Genome Jungle.* London: Arum, 1994.

Corea, Gena. 'Women, Class & Genetic Engineering. The Effects of New Reproductive Technologies on All Women.' *The Baby Machine. Reproductive Technology and the Commercialisation of Motherhood.* Ed. Jocelynne E. Scutt. London: Merlin, 1988. 135–56.

Corn, Joseph J. *Imagining Tomorrow. History, Technology, and the American Future.* Cambridge: MIT Press, 1986.

Cranor, Carl F. (ed.). *Are Genes Us? The Social Consequences of the New Genetics*. New Brunswick: Rutgers University Press,1994.

Crary, Jonathan and Sanford Kwinter (eds). *Incorporations*. New York: Zone, 1992.

Crawford, T. Hugh. 'Imaging the Human Body: Quasi Objects, Quasi Texts, and the Theater of Proof.' *PMLA* 3.1 (January 1996): 66–79.

Crichton, Michael. *The Andromeda Strain*. New York: Knopf, 1969.

Davis, Joel. *Mapping the Code. The Human Genome Project and the Choices of Modern Science*. New York: Wiley, 1990.

Davis, Kathy. 'Towards a Feminist Rhetoric: The Gilligan Debate Revisited.' *Women's Studies International Forum* 15.2 (1992): 219–31.

Dawkins, Richard. *The Selfish Gene*. Oxford: Oxford University Press, 1989. Originally published 1976.

De Kruif, Paul. *Microbe Hunters*. New York: Harcourt, 1926.

De Lauretis, Teresa (ed.). *Feminist Studies/Critical Studies*. Bloomington: Indiana University Press, 1986.

Dickstein, Morris. *Gates of Eden. American Culture in the Sixties*. New York: Penguin, 1977.

Doyle, Richard. 'Vital Language.' *Are Genes Us? The Social Consequences of the New Genetics*. Ed. Carl F. Cranor. New Brunswick: Rutgers University Press, 1994, 52–68.

Durant, J., G. Evans and G. Thomas. 'Public Understanding of Science.' *Nature* 340 (1989): 11–14.

Durant, John, Anders Hansen and Martin Bauer, 'Public Understanding of the New Genetics.' *The Troubled Helix. Social and Psychological Implications of the New Human Genetics*. Eds Theresa Marteau and Martin Richards. Cambridge: Cambridge University Press, 1996, 235–48.

Duster, Troy. *Backdoor to Eugenics*. New York: Routledge, 1990.

Dutton, Diana B. and Nancy E. Pfund. *Worse than Disease. Pitfalls of Medical Progress*. Cambridge: Cambridge University Press, 1988.

Elkington, John. *The Gene Factory. Inside the Genetic and Biotechnology Business Revolution*. New York: Caroll and Graff, 1985.

Erni, John. 'Articulating the (Im)possible: Popular Media and the Cultural Politics of "Curing Aids".' *Communication* 13 (1992): 39–56.

Escobar, Arthur. 'Welcome to Cyberia. Notes on the Anthropology of Cyberculture.' *Current Anthropology* 35.3 (June 1994): 211–31.

Etzioni, Amitai. *The Genetic Fix. The Next Technological Revolution*. New York: Harper, 1973.

Favret, Mary A. 'A Woman Writes the Fiction of Science: The Body in Frankenstein.' *Genders* 14 (Fall 1992): 50–65.

Firestone, Shulamith. *The Dialectic of Sex. The Case for a Feminist Revolution*. New York: Bantam, 1970.

Fletcher, Joseph. *The Ethics of Genetic Control. Ending Reproductive Roulette*. New York: Anchor, 1974.

Fogle, Thomas. 'Information Metaphors and the Human Genome Project.' *Perspectives in Biology and Medicine* 38.4 (Summer 1995): 535–53.

Folette, Marcel C. *Making Science our Own. Public Images of Science 1910–1950*. Chicago: Chicago University Press, 1990.

Foucault, Michel. *The Archaeology of Knowledge.* New York: Pantheon, 1972. Originally published 1969.

Francoeur, Robert T. *Utopian Motherhood. New Trends in Human Reproduction.* New York: Barnes, 1970.

Franklin, Sarah B. 'Life Story. The Gene as Fetish and Object on TV.' *Science as Culture* 12 (1988): 92–100.

Freedman, Nancy. *Joshua, Son of None.* London: Granada, 1973.

Gilbert, Walter. 'A Vision of the Grail.' *The Code of Codes. Scientific and Social Issues in the Human Genome Project.* Eds Daniel J. Kevles and Leroy Hood. Cambridge: Harvard University Press, 1992, 83–97.

Gitlin, Todd. *The Whole World is Watching. Mass Media and the Making and Unmaking of the New Left.* Berkeley: University of California Press, 1980.

Goldsmith, Maurice. *The Science Critic. A Critical Analysis of the Popular Presentation of Science.* London: Routledge, 1986.

Goodfield, June. *Playing God. Genetic Engineering and the Manipulation of Life.* London: Hutchinson, 1977.

Goodfield, June. *Reflections on Science and Media.* Washington: American Association for the Advancement of Science, 1981.

Gould, Stephen Jay. *Ever Since Darwin. Reflections on Natural History.* New York: Norton, 1977.

Green, Jeremy. 'Media Sensation and Science. The Case of the Criminal Chromosome.' *Expository Science. Forms and Functions of Popularisation.* Eds Terry Shinn and Richard Whitley. Dordrecht: Reidel, 1985, 139–61.

Gross, Alan G. *The Rhetoric of Science.* Cambridge: Harvard University Press, 1990.

Gusfield, Joseph. *Drinking-Driving and the Symbolic Order.* Chicago: University of Chicago Press, 1981.

Halacy, D.S. Jr. *Cyborg. Evolution of the Superman.* New York: Harper and Row, 1965.

Hall, Stephen. *Invisible Frontiers. The Race to Synthesize a Human Gene.* Washington: Microsoft Press, 1987.

Haller, Mark H. *Myth of the Menace of the Feebleminded in Eugenics. Hereditarian Attitudes in American Thought.* New Brunswick: Rutgers University Press, 1963.

Hallin, Daniel C. *The Uncensored War. The Media and Vietnam.* Berkeley: University of California Press, 1986.

Hamilton, M. (ed). *The New Genetics and the Future of Man.* Grand Rapids: Eerdmans, 1972.

Haraway, Donna. 'Situated Knowledges: The Science Question in Feminism and the Privilege of the Partial Perspective.' *Feminist Studies* 14:3 (Fall 1988): 575–99.

Haraway, Donna. 'The Biopolitics of Postmodern Bodies: Determinations of Self in Immune Discourse.' *Difference* 1 (Winter 1989): 3–43.

Haraway, Donna. *Primate Visions. Gender, Race and Nature in the World of Modern Science.* New York: Routledge, 1989.

Haraway, Donna. *Simians, Cyborgs and Women. The Reinvention of Nature.* London: Free Association Books, 1991.

Harré, Rom. 'Some Narrative Conventions of Scientific Discourse.' *Narrative in Culture. The Uses of Storytelling in the Sciences, Philosophy, and Literature.* Ed. Christopher Nash. London: Routledge, 1990, 102–26.

Hartouni, Valerie. 'Breached Births: Reflections on Race, Gender and Reproductive Discourse in the 1980s.' *Configurations* 1 (1994): 73–88.

Harvey, David. *The Condition of Postmodernity*. Oxford: Basil Blackwell, 1989.

Hayles, N. Katherine. 'The Materiality of Informatics.' *Configurations* 1 (1992): 147–70.

Hayles, N. Katherine. 'Constrained Constructivism. Locating Scientific Inquiry in the Theater of Representation.' *Realism and Representation*. Ed. George Levine. Madison: University of Wisconsin Press, 1993, 27–43.

Haynes, Rosalynne D. *From Faust to Strangelove. Representations of the Scientist in Western Literature*. Baltimore: Johns Hopkins University Press, 1994.

Hellmann, John. *Fables of Fact. The New Journalism as New Fiction*. Urbana: University of Illinois Press, 1980.

Herbert, Frank. *The Eyes of Heisenberg*. New York: Berkley, 1966.

Hesse, Mary B. and Max Black. *Models and Analogies in Science*. Notre Dame, Indiana: Notre Dame University Press, 1970.

Hilgartner, Stephen. 'The Dominant View of Popularization. Conceptual Problems, Political Uses.' *Social Studies of Science* 20 (1990): 519–39.

Hill, Stephen and Tim Turpin. 'Academic Research Cultures in Collision.' *Science as Culture* 20 (1994): 327–61.

Hilton, B. *et al.* (eds). *Ethical Issues in Human Genetics*. New York: Plenum, 1973.

Hubbard, Ruth and Elijah Wald. *Exploding the Gene Myth. How Genetic Information is Produced and Manipulated by Scientists, Physicians, Employers, Insurance Companies, Educators, and Law Enforcers*. Boston: Beacon, 1993.

Hutton, Richard. *Bio-Revolution. DNA and the Ethics of Man-Made Life*. New York: New American Library, 1978.

Huxley, Aldous. *Brave New World*. New York: Harper and Row, 1946. Originally published 1932.

Huxley, Aldous. *Brave New World Revisited*. New York: Harper and Row, 1958.

Jacobi, Daniel and Bernard Schiele. 'Scientific Imagery and Popularized Imagery. Differences and Similarities in the Photographic Portraits of Scientists.' *Social Studies of Science* 19 (1989): 731–53.

Jameson, Fredric. 'Periodizing the Sixties.' *The Sixties without Apology*. Eds Sohnya Sayres *et al.* Minneapolis: University of Minnesota Press, 1984. 178–209.

Jameson, Fredric. 'Postmodernisms, or the Logic of Late Capitalism.' *New Left Review* 146 (1984): 53–92.

Judson, Horace F. *The Eighth Day of Creation: Makers of the Revolution in Biology*. New York: Simon and Schuster, 1978.

Keller, Evelyn Fox. 'Making Gender Visible in the Pursuit of Nature's Secret.' *Feminist Studies/Critical Studies*. Ed. Teresa de Lauretis. Bloomington: Indiana University Press, 1986, 67–77.

Keller, Evelyn Fox. *Secrets of Life, Secrets of Death. Essays on Language, Gender, and Science*. New York: Routledge, 1992.

Keller, Evelyn Fox. 'Master Molecules.' *Are Genes Us? The Social Consequences of the New Genetics*. Ed. Carl F. Cranor. New Brunswick: Rutgers University Press, 1994.

Keller, Evelyn Fox. *Refiguring Life. Metaphors of Twentieth-Century Biology*. New York: Columbia University Press, 1995.

Kenney, Martin. *Biotechnology. The University-Industrial Complex*. New Haven: Yale University Press, 1986.

Kevles, Daniel. *In the Name of Eugenics. Genetics and the Uses of Human Heredity*. Berkeley: University of California Press, 1985.

Kevles, Daniel J. and Leroy Hood (eds). *The Code of Codes. Scientific and Social Issues in the Human Genome Project*. Cambridge: Harvard University Press, 1992.

Klein, Renate D. 'Genetic and Reproductive Engineering. The Global View.' *The Baby Machine. Reproductive Technology and the Commercialisation of Motherhood*. Ed. Jocelynne E. Scutt. London: Merlin, 1988, 235–72.

Knight, Nancy. '"The New Light". X-rays and Medical Futurism.' *Imagining Tomorrow. History, Technology, and the American Future*. Ed. Joseph J. Corn. Cambridge: MIT Press, 1986, 10–34.

Krimsky, Sheldon. *Genetic Alchemy: The Social History of the Recombinant DNA Controversy*. Cambridge: MIT Press, 1982.

Krimsky, Sheldon. *Biotechnics and Society. The Rise of Industrial Genetics*. New York: Praeger, 1991.

Kronberg, Arthur. *The Golden Helix. Inside Biotech Ventures*. Sausalito: University Science Books, 1995.

Labinger, Jay. 'Encoding an Infinite Message: Richard Powers's *The Gold Bug Variations*.' *Configurations* 3 (1995): 79–94.

Lakoff, George and Mark Johnson. *Metaphors We Live By*. Chicago: University of Chicago Press, 1980.

Latour, Bruno. *Science in Action. How to Follow Scientists and Engineers through Society*. Cambridge: Harvard University Press, 1987.

Latour, Bruno. 'Drawing Things Together.' *Representation in Scientific Practice*. Eds. Steve Woolgar and Michael Lynch. Boston: MIT Press, 1990, 19–67.

Latour, Bruno. *We Have Never Been Modern*. Cambridge: Harvard University Press, 1993.

Latour, Bruno and Steve Woolgar. *Laboratory Life. The Social Construction of Scientific Facts*. London: Sage, 1979.

Lavallay, Albert J. 'The Stage and Film Children of Frankenstein.' *The Endurance of Frankenstein*. Eds George Levine and U.C. Knoepflmacher. Berkeley: University of California Press, 1979, 243–88.

Lear, John. *Recombinant DNA. The Untold Story*. New York: Crown, 1978.

Levin, Ira. *The Boys from Brazil*. London: Pan, 1976.

Levine, George (ed.). *Realism and Representation*. Madison: University of Wisconsin Press, 1993.

Levine, George and U.C. Knoepflmacher (eds). *The Endurance of Frankenstein*. Berkeley: University of California Press, 1979.

Lewenstein, Bruce. 'The Meaning of Public Understanding of Science in the US after World War II.' *Public Understanding of Science* 1 (1992): 45–68.

Lewontin, R.C. *Biology as Ideology. The Doctrine of DNA*. Concord, Ontario: Anansi, 1991.

Lewontin, R.C. 'The Dream of the Human Genome.' *New York Review of Books* (28 May 1992): 31–9.

Lewontin, R.C., Steven Rose and Leon Kamin. *Not in our Genes. Biology, Ideology, and Human Nature*. New York: Pantheon, 1984.

Locke, David. *Science as Writing*. New Haven: Yale University Press, 1992.

Lyotard, Jean-François. *The Postmodern Condition: A Report on Knowledge. 1979*. Manchester: Manchester University Press, 1984.

Maassen, Sabine and Peter Weingart. 'Metaphors – Messengers of Meaning. A Contribution to an Evolutionary Sociology of Science.' *Science Communication* 17 (1995): 9–31.

Marteau, Theresa and Martin Richards (eds). *The Troubled Helix. Social and Psychological Implications of the New Human Genetics*. Cambridge: Cambridge University Press, 1996.

Martin, Emily. *The Woman in the Body. A Cultural Analysis of Reproduction*. Boston: Beacon, 1987.

Massey, Doreen, Paul Quintas and David Wield. *High-Tech Fantasies. Science Parks in Society, Science, and Space*. London: Routledge, 1992.

Mattelart, Armand and Michele Mattelart. *Rethinking Media Theory. Signposts and New Directions*. Minneapolis: University of Minnesota Press, 1992.

McCormack, Thelma (ed.). *News and Knowledge*. London, Conn.: Jai Press, 1986.

Mellor, Anne K. *Mary Shelley. Her Life, her Fiction, her Monsters*. New York: Routledge, 1988.

Mitchison, Naomi. *Solution Three*. London: Dobson, 1975.

Moers, Ellen. 'Female Gothic.' *The Endurance of Frankenstein*. Eds George Levine and U.C. Knoepflmacher. Berkeley: University of California Press, 1979, 77–87.

Montgomery, Scott L. 'Codes and Combat in Biomedical Discourse.' *Science as Culture* 2.12 (1991): 341–90.

Muller, Hermann J. *Out of the Night. A Biologist's View of the Future*. New York: Vanguard Press, 1935.

Myers, Greg. 'Making a Discovery. Narratives of Split Genes.' *Narrative in Culture. The Uses of Storytelling in the Sciences, Philosophy, and Literature*. Ed. Christopher Nash. London: Routledge, 1990, 102–26.

Myers, Greg. *Writing Biology. Texts in the Social Construction of Scientific Knowledge*. Madison: University of Wisconsin Press, 1990.

Nash, Christopher (ed.) *Narrative in Culture. The Uses of Storytelling in the Sciences, Philosophy, and Literature*. London: Routledge, 1990.

Nelkin, Dorothy. *Selling Science. How the Press Covers Science and Technology*. New York: Freeman, 1987.

Nelkin, Dorothy and Laurence Tancredi. *Dangerous Diagnostics. The Social Power of Biological Information*. New York: Basic Books, 1989.

Nelkin, Dorothy and M. Susan Lindee. *The DNA Mystique. The Gene as a Cultural Icon*. New York: Freeman, 1995.

Nossal, G.J.V. and Ross L. Coppal. *Reshaping Life. Key Issues in Genetic Engineering*. Cambridge: Cambridge University Press, 1989.

Olby, Robert. *The Path to the Double Helix*. London: Macmillan, 1974.

Ormiston, Gayle L. and Raphael Sassower. *Narrative Experiments. The Discursive Authority of Science and Technology*. Minneapolis: University of Minnesota Press, 1989.

Pauly, John. 'The Politics of New Journalism.' *Literary Journalism in the Twentieth Century*. Ed. Norman Sims. Oxford: Oxford University Press, 1990, 110–30.

Peterfreund, Stuart (ed.). *Literature and Science. Theory and Practice*. Boston: Northeastern University Press, 1990.

Piercy, Marge. *Woman on the Edge of Time*. New York: Fawcett Crest, 1976.

Pollack, Robert. *Signs of Life. The Language and Meanings of DNA*. New York: Houghton Miffkin, 1994.

Poster, Mark. *The Mode of Information. Poststructuralism and Social Context*. Chicago: University of Chicago Press, 1990.

Powers, Richard. *The Gold Bug Variations*. New York: Harper, 1991.

Rabinow, Paul. 'Artificiality and Enlightenment. From Sociobiology to Biosociality.' *Incorporations*. Eds. Jonathan Crary and Sanford Kwinter. New York: Zone, 1992, 234–52.

Rabinow, Paul. *Making PCR. A Story of Biotechnology*. Chicago: University of Chicago Press, 1996.

Ramsey, Paul. *Fabricated Man. The Ethics of Genetic Control*. New Haven: Yale University Press, 1970.

Raymond, Janice G. *Women as Wombs. Reproductive Technologies and the Battle over Women's Freedom*. New York: Harper, 1993.

Rifkin, Jeremy. *Algeny*. New York: Viking, 1983.

Rogers, Michael. *Biohazard*. New York: Knopf, 1977.

Rorvik, David M. *In his Image. The Cloning of a Man*. New York: Lippincott, 1978.

Rose, Hillary. *Love, Power, and Knowledge. Towards a Feminist Transformation of the Sciences*. Bloomington: Indiana University Press, 1994.

Rosner, Mary and T.R. Johnson. 'Telling Stories: Metaphors of the Human Genome Project.' *Hypathia* 10:4 (Fall 1995): 104–29.

Rouse, Joseph. 'What Are the Cultural Studies of Scientific Knowledge?' *Configurations* 1 (1992): 1–22.

Rowland, Robyn. *Living Laboratories. Women and Reproductive Technologies*. Bloomington: Indiana University Press, 1992.

Sargent, Pamela. *Cloned Lives*. New York: Fawcett, 1976.

Sayre, Anne. *Rosalind Franklin and DNA. The First Full Account of Rosalind Franklin's Role in the Discovery of the Double Helical Structure of DNA*. New York: Norton, 1975.

Sayres, Sohnya *et al.* (eds). *The Sixties without Apology*. Minneapolis: University of Minnesota Press, 1984.

Schrödinger, Erwin. *What is Life? The Physical Aspects of the Living Cell*. Cambridge: Cambridge University Press, 1967. Originally published 1944.

Scutt, Jocelynne E. (ed.). *The Baby Machine. Reproductive Technology and the Commercialisation of Motherhood*. London: Merlin, 1988.

Segal, Howard P. *Future Imperfect. The Mixed Blessings of Technology in America*. Boston: University of Massachusetts Press, 1994.

Shelley, Mary. *Frankenstein or the Modern Prometheus*. London: Penguin Books, 1985. Originally published 1818.

Sherwin, Susan. 'Feminist and Medical Ethics: Two Different Approaches to Contextual Ethics.' *Hypathia* 4:2 (Summer 1989): 17–29.

Shinn, Terry and Richard Whitley (eds). *Expository Science. Forms and Functions of Popularisation*. Dordrecht: Reidel, 1985.

Sismondo, Sergio. *Science without Myth. On Constructions, Reality, and Social Knowledge*. New York: SUNY Press, 1996.

Spallone, Patricia. *Beyond Conception. The New Politics of Reproduction*. Granby: Bergin, 1989.

Spanier, Bonnie. *Im/partial Science. Gender Ideology in Molecular Biology.* Bloomington: Indiana University Press, 1995.

Squier, Susan Merrill. *Babies in Bottles. Twentieth-Century Visions of Reproductive Technology.* New Brunswick: Rutgers University Press, 1994.

Stafford, Barbara. *Body Criticism. Imaging the Unseen in Enlightenment Art and Medicine.* New York: Zone, 1991.

Stone, Allucquere R. 'Will the Real Body Please Stand Up? Boundary Stories about Virtual Culture.' *Cyberspace. First Steps.* Ed. Michael Benedikt. Cambridge: MIT Press, 1991): 81–118.

Strathern, Marilyn. *Reproducing the Future. Anthropology, Kinship, and the New Reproductive Technologies.* Manchester: Manchester University Press, 1993.

Suzuki, David and Peter Knudtson. *Genethics. The Clash between the New Genetics and Human Values.* Cambridge: Harvard University Press, 1990.

Taylor, Charles A. *Defining Science. A Rhetoric of Demarcation.* Madison: University of Wisconsin Press, 1996.

Taylor, Gordon Rattray. *The Biological Time Bomb.* New York: World Publishing, 1968.

Teitelman, Robert. *Gene Dreams. Wall Street, Academia, and the Rise of Biotechnology.* New York: Harper, 1989.

Thompson, Larry. *Correcting the Code. Inventing the Genetic Cure for the Human Body.* New York: Simon and Schuster, 1994.

Thomson, Amy. *Virtual Girl.* New York: Ace, 1993.

Thornburg, Mary K. Patterson. *The Monster in the Mirror.* Ann Arbor: University of Michigan Press, 1987.

Tuchman, Gaye. *Making News. A Study in the Construction of Reality.* New York: Free Press, 1978.

Turney, Jon. 'Thinking about the Human Genome Project' *Science as Culture* 19 (1993): 282–94.

Van Dijck, José. *Manufacturing Babies and Public Consent. Debating the New Reproductive Technologies.* London: Macmillan, 1995.

Van Dijck, José. 'The Human Genome Narrative.' *Science as Culture* 23 (1995): 217–47.

Van Lente, Harro. *Promising Technology. The Dynamics and Expectations in Technological Developments.* Delft: Eburon, 1993.

Warren, Virginia L. 'Feminist Directions in Medical Ethics.' *Hypathia* 4:2 (Summer 1989): 31–45.

Watson, James D. *The Double Helix.* New York: Mentor, 1968.

Watson, James D. and John Tooze. *The DNA Story. A Documentary History of Gene Cloning.* San Francisco: Freeman, 1981.

Weber, Ronald. *The Literature of Fact. Literary Nonfiction in American Writing.* Athens: Ohio State University Press, 1980.

Weeks, Jeffrey. *Sexuality and its Discontents. Meanings, Myths and Modern Sexualities.* London: Routledge, 1985.

Wells, H.G. *The Island of Dr Moreau.* New York: New American Library, 1977. Originally published 1896.

Wilhelm, Kate. *Where Late the Sweet Birds Sang.* New York: Harper, 1976.

Williams, Rosalind. *Notes on the Underground. An Essay in Technology, Society and the Imagination.* Cambridge: MIT Press, 1990.

Wilson, Edward O. *Sociobiology. The New Synthesis*. Cambridge: Harvard University Press, 1975.

Wingerson, Lois. *Mapping our Genes: The Genome Project and the Future of Medicine*. New York: Penguin, 1990.

Wolfe, Tom and E.W. Johnson (eds). *The New Journalism*. New York: Picador, 1974.

Woolgar, Steve and Michael Lynch. *Representation in Scientific Practice*. Boston: MIT Press, 1990.

Woolmark, Jenny. *Aliens and Others. Science Fiction, Feminism, and Postmodernism*. Iowa City: Iowa University Press, 1994.

Wynne, Brian. 'Public Understanding of Science Research: New Horizons or Hall of Mirrors?' *Public Understanding of Science* 1 (1992): 39–43.

Yoxen, Edward. *The Gene Business. Who Should Control Biotechnology?* New York: Harper, 1983.

Yoxen, Edward. 'Speaking about Competition. An Essay on the Double Helix as Popularisation.' *Expository Science. Forms and Functions of Popularisation*. Eds Terry Shinn and Richard Whitley. Dordrecht: Reidel, 1985.

Zelizer, Barbie. *Covering the Body. The Kennedy Assassination, the Media, and the Shaping of Collective Memory*. Chicago: University of Chicago Press, 1992.

Zilinskas, Raymond A. and Burke K. Zimmerman (eds). *The Gene-Splicing Wars*. New York: Macmillan, 1986.

Zinder, Norton D. 'A Personal View of the Media's Role in the Recombinant DNA War.' *The Gene-Splicing Wars*. Eds Raymond A. Zilinskas and Burke K. Zimmerman. New York: Macmillan, 1986.

Index